Heinz Blatter, Thomas Greber

Radical Relativity

Heinz Blatter, Thomas Greber

Radical Relativity

An Uncommon Way to Spacetime

DE GRUYTER

Authors
Prof. Dr. Heinz Blatter
Luzernerstrasse 15
CH-4800 Zofingen
Switzerland
blatter@env.ethz.ch

Prof. Dr. Thomas Greber
Physik-Institut
University of Zurich
Winterthurerstrasse 190
CH-8057 Zurich
Switzerland
greber@physik.uzh.ch

ISBN 978-3-11-150309-7
e-ISBN (PDF) 978-3-11-150359-2
e-ISBN (EPUB) 978-3-11-150366-0

Library of Congress Control Number: 2024952713

Bibliographic information published by the Deutsche Nationalbibliothek
The Deutsche Nationalbibliothek lists this publication in the Deutsche Nationalbibliografie; detailed bibliographic data are available on the Internet at http://dnb.dnb.de.

Cover image: "Der Weltraumfahrer (The space traveller)" by Karl Blatter, Switzerland (1916-1996), around

www.degruyter.com
Questions about General Product Safety
Regulation: productsafety@degruyterbrill.com

Preface

It was an eye-opening encounter in the evening train commuting home from Zurich back in the early 1980ies. I (HB) was studying some book on fluid dynamics to prepare a lecture. A young native English speaking couple sitting opposite in the same compartment became curious about what I am reading. When they realized that this was a physics book they started enthusiastically telling me that they were just reading George Gamow's book "Mister Tompkins in Wonderland" (Gamow, 1965). They showed me the picture with Tompkins cycling with relativistic speed through a city street canyon showing length contracted houses as they were supposed to look for Mr. Tompkins. I explained them that this is not what Mr. Tompkins sees from the bicycle, but corresponds to a different type of observation. I then explained them aberration and what Tompkins actually would have seen. I was truly surprised when I realized their reaction: they were extremely disappointed and almost angry, not at me, but at the fact that not even such accepted books by eminent scientists were reliable. What could they rely on if not on such information?

The confusion about length contraction and aberration prevailed through a long time after the Einstein (1905a) publication although Einstein made a distinction between the two conceps of "observers". The confusion perhaps started because Einstein used the German word *Beobachter* (observer) for the two different viewpoints, the local observer looking around from his position in space and the *geometry* of spacetime with points, distances, and synchronized clocks. Although Lampa (1924) correctly described the appearance of moving bodies to a local observer (in German language), the erroneous description of length contracted appearance of bodies unfortunately remained a common habit. More than 30 years later, Terrell (1959) reinterpreted the appearance in an approximation for very small bodies as a rotated appearance. This interpretation again stirred some confusion of what exactly is rotated: the entire bicycle or the license plate and wheels separately, as can be argued easily (Gamow, 1961). This pragmatic interpretation, pretending simplicity, also obscures the beauty of the symmetries behind aberration. Penrose (1959) exploited these symmetries by presenting four proofs on a few lines that a spherical object always presents a circular outline to any local observer independent of his motion. The potential in the patterns of aberration to serve as a foundation of special relativity was presented in a short but stimulating article by Komar (1965). This article became the starting point to our work on teaching relativity in a more intuitive but still rigorous way, which we tried to condense in the present work.

We asked several colleges from theoretical physics to review the work and give us their opinion. Sometimes, they immediately asked back why we don't write all in covariant form where all is much *easier* and clearer. This may be correct for physicists, and these books are already written, but it is not useful for the target audience we have in mind. The theory of special relativity is perhaps the simplest non-trivial theory reaching beyond our picture of the world, which is limited by our senses and everyday experi-

https://doi.org/10.1515/9783111503592-202

ence. This makes the theory an attractive example to teach physics as more than a set of equations and technical procedures but to demonstrate physical thinking.

Radical relativity is a set of ideas about teaching special relativity in an intuitive but rigorous way. To reach this goal, several patterns of standard introductions are omitted: no history, no synchronization, no paradoxes. The conclusions are the same, however, a terminology radically adjusted to the theory reveals some surprising aspects and insights.

Relativistic spacetime requires a revision of classical dynamics. If Newtons second law is written in quantities that are defined in the frame of reference of the moving body on which the force is acting the introduction of a relativistic mass can be avoided. This is used to establish relativistic *dynamics* in a comprehensive way that backs on the *kinematic* consequences of Special Relativity alone.

The special theory of relativity is a theory of space and time, and as such, at the very foundation of physics. Everyday experience gives us the impression of having a clear picture of what is space and time as two independent phenomena. Everyday experience is limited to a narrow range of distances, time spans, and especially velocities. Within these ranges the classical view of space and time is valid to an extent that can not be challenged by everyday observations. On the other hand, phenomena such as light, radio waves, electricity, and magnetism, although they belong to our everyday life, can only be properly understood as consequences of qualities of space and time that become relevant far outside the range of personal perception.

The first comprehensive description of electrodynamics developed through the 1860ies and presented by Maxwell (1873) was the first genuine relativistic theory. The wave solution was already found by Maxwell, but only verified experimentally some 20 years later by Hertz (1888b,a).

Perhaps one of the crucial points in the theory of Lorentz (1895) was the difference in the velocity addition theorem between electrodynamics and mechanics (Einstein, 1982), which was resolved in the Einstein (1905a) paper on "Die Elektrodynamik bewegter Körper". At almost the same time, Poincaré (1905) published a short paper "Sur la dynamique de l'électron" and a year later a comprehensive work with the same title (Poincaré, 1906). Minkowskis mathematical formulation of spacetime radically changed the classical picture of space and time as independent entities and showed that space and time must be treated mathematically as a "union" (Minkowski, 1909).

This book intends to present Special Relativity to students and physics teachers, however, not in the sense of a textbook covering all the theoretical and technical aspects of the theory in full depth. The challenge is to teach relativity in a way that a deeper understanding is possible without coverage of all aspects in every detail. In this effort we follow a recommendation, that proved to be extremely helpful, of late Markus Fierz, a professor of theoretical physics at ETH Zurich, Switzerland:

> If you want to learn a new field (in physics), then try to understand the simplest non-trivial example.

Usually such examples contain all aspects of a field that are relevant for the understanding. Generality and completeness are certainly relevant aspects for professional research and applications of a scientific field, however, not necessarily for teaching and the students understanding. The principal aims of the text are illustrated by the following seven points:
1. The different types of observers, a *local individual observer* looking at the world around him from one point in space (and time) and the *Einsteinian observer* who represents the description of events in spacetime. The first is called *observer*, the latter *frame of reference* in the present text. The confusion of these two types of observers caused some pitfalls in relativistic texts in the past (Section 1.3.1).
2. Different measures of the speed of a motion are strictly distinguished depending on the frame of reference, in which the distance is measured, and the frame of reference, in which the time is measured. The different possibilities, called *coordinate velocity* (or velocity), *proper velocity* (or celerity) and *rapidity* are useful for derivation of the various kinematic and dynamic relations and laws (Section 3.1).
3. Rapidity is measured entirely within the frame of the accelerated object and thus links spacetime and kinematics to inertia and forces (Section 3.1.2)
4. Electrodynamics is genuinely relativistic and has no classical counterpart. This makes special relativity ubiquitous (omnipresent) in daily life, although in an abstract and not necessarily intuitive way (Section 3.3).
5. A hierarchical construction of physics is suggested, starting with space and geometry and next with time and kinematics i.e. moving objects. The objects can be "charged" with additional properties such as *inertial mass* leading to dynamics with forces and acceleration, with *electric charge*, leading to electrodynamics, and with *gravitational mass* leading to gravity (Section 1.4.1).
6. Physics has aspects resembling the axiomatic approach in mathematics and can be based on a comparably small number of *hypotheses*. Space, time, inertia, and electric charge are introduced through few hypotheses, and we illustrate how mathematics finally restricts the possibilities for the formulation of physics, making the axiomatic approach to physics particularly powerful (Section 1.4).
7. The importance and presence of relativistic phenomena in daily life, science and science fiction is used to demonstrate the specific relativistic phenomena in comparison with their classical counterparts in daily life and technology (Chapter 6) and science fiction by discussing the possibilities and limits of an accelerated intergalactic spaceflight (Chapter 7).

The project "Radical Relativity" emerged from many discussions that we lead during lunch on the many Thursday's when we found time for *discorsos*. It does not cover everything we talked about, but it is a topic we visited frequently in our conversations.

Our first *acknowledgments* goes to a class in the Gymnasium (College) in Zofingen, Switzerland, where one of us (HB) was teaching mathematics and physics back in the 1970ies. The students asked me to teach them special relativity, a request that I gladly followed. I chose a few textbooks on relativity on the appropriate level and followed their contents relatively closely. However, I had bad feelings in some parts when explaining verbally what we just learnt in some mathematical form. Driven by this I spent days in the library of ETH Zurich searching the literature on relativity, and, my bad feelings were confirmed. I encountered the Penrose (1959) paper on *The apparent shape of a relativistically moving sphere* which identified the confusion between length contraction and aberration common in most textbooks of that time. Thus, the unusual request of the Gymnasium class finally triggered my interest and the subsequent studies about didactics of relativity. I also acknowledge the unknown couple on the train that opened my eyes to the responsibility when teaching physics. Of course, it is always the contemporarily accepted wisdom that is presented in textbooks and most likely, it will eventually be corrected in details in the future. It is now a long time since the request of the students triggered the start of this work. Perhaps, such a book is never truly finished, but eventually one must decide to expose it.

Zofingen and Zürich,
December 2023

Heinz Blatter
Thomas Greber

Contents

Prologue

Any physical theory that involves motion relies on a concept of space and time. Mechanics bases on the Newtonian principles and electrodynamics on the Maxwell equations. Both theories are abstractions of everyday experiences in different aspects. Newtonian mechanics are more accessible: it is relatively easy to convince "somebody" that a pendulum or a clock has the same frequency on a boat that moves down a quiet river, as an identical clock on the riverside. Maxwellian electrodynamics on the other hand is less intuitive: It implies that the speed of light is the same, no matter whether the light is emitted from a lantern on the boat or on the riverside. Therefore, the theory of Newton and that of Maxwell are in conflict. The inconsistency emerges from two different concepts of space and time. Newton's theory obeys the classical picture of independent space and time while Maxwell's theory obeys special relativity in spacetime, where space and time are coupled. This has, for example, intriguing consequences if the velocity of the boat relative to the riverside shall be determined. A measurement of velocity requires the determination of a time interval during which a certain path is crossed. In contrast to classical mechanics, in special relativity it has to be distinguished whether the path length and the time to cross it are both measured on the riverside or the boat, or whether they are not measured in the same frame of reference. Reassuringly, the difference between the two velocities disappears for small speeds compared to the speed of light. All phenomena related to electricity and magnetism, are intimately related to relativity. Phenomena such as light, electric motors, reception of radio and TV programs, Wi-Fi wireless networks, or magnetic induction heating in the kitchen, are a consequence of relativistic processes and do not appear as "natural" as mechanical motion does. Still, special relativity also affects mechanics, and we should be aware of its consequences if we are interested in a consistent physical view of the universe.

In the first Chapter, we outline the concept of spacetime. The relevant hypotheses about empty space and time form the basis of our view, which is a radical approach to relativity that starts from geometry and chronology. In the second Chapter, we develop the theory of special relativity from a local and a non-local point of view. In the third Chapter, we work out the consequences of special relativity, for example, the distinction between different speeds. In the fourth Chapter seeming contradictions are revisited. In the fifth Chapter, we outline the limits of special relativity and touch general relativity. In the sixth Chapter, we discuss applications like synchrotron radiation. Finally, the last Chapter explores relativistic space travel.

This book is a monograph on the special relativity theory starting from the view of a local observer. It is an uncommon way to spacetime with rarely found or skipped aspects, and does not replace textbooks, but complements them.

https://doi.org/10.1515/9783111503592-001

1 The concept of Spacetime

The idea of connecting and ordering events requires a concept of space and time. It is at the foundation of any physical theory on any process. This Chapter discusses the underlying hypotheses of space and time and their unification into spacetime involving a maximum speed. Finally, more hypotheses like those of mass, charge, spin, and that of the wave-particle duality are outlined.

1.1 Hypotheses

The strength of science in general, and physics in particular, rests upon the reliability of deductions derived from *hypotheses*. A scientific hypothesis is a fundamental assumption that cannot be proved but can be tested and it is the starting point of any theory which is a structure that enables understanding and predictions. If a hypothesis like the claim that space and time are independent is falsified this has implications for the theories that are built on such hypotheses.

Our world exhibits a large variety of phenomena and our brain works in a way that allows us to structure this multitude. Accordingly, the first stage in the development of science is the collection of information and its empirical classification, before hypotheses may be formulated and before a theory can be developed. Perhaps, the ultimate goal of scientific theorizing is the reduction of the underlying assumptions to a minimal number of hypotheses of utmost simplicity on which a unified theory may be built.

This *reductionism* is sometimes criticized and disqualified as unnatural and narrow-minded, although bringing structure and order into the perception of the world is essential for survival. Mathematics demonstrates the power of rigorous deduction from underlying systems of axioms. It is an astonishing experience to realize the extent to which natural (physical) phenomena can be described precisely with mathematical language. This also motivates theoreticians to map the powerful tools of mathematics onto the realm of physics, or the other way round, to map physical ideas onto mathematical structures. The resulting set of hypotheses should therefore fulfill similar criteria as the systems of axioms for mathematical fields: consistency, completeness, and independence. A system of physical hypotheses must meet additional requirements, such as consistency with observations, but it should also be intuitive, convincing, and simple, although these requirements are more a matter of taste and aesthetics rather than a matter of physics itself.

https://doi.org/10.1515/9783111503592-002

1.2 Space and time, spacetime

Almost every human and perhaps some animals have an intuitive understanding of space and time. Space hosts objects and time passes. Thus, hypotheses concerning space and time are convincing if they correspond to this intuitive picture.

The hypotheses on space and time assume basic symmetries such as the isotropy and homogeneity of space and time, which often are taken for granted: why should the world work differently at another place or another time?

Hypotheses on space include its continuous extension in three dimensions where three independent coordinates are used to describe positions of objects. With metrics, it is possible to measure unambiguously distances between different points and angles between different directions. Euclidian Geometry is the corresponding theory in this flat three-dimensional space and is one of the hypotheses of special relativity too. The concept of curved space emerges in general relativity, where distances between points become ambiguous and depend on the arrangement of masses and energy in non-empty space.

Hypotheses on time include a continuous extension in one dimension where one coordinate is used to describe moments or points in time. The metrics allow to determine time intervals and to order sequences of moments. The special theory of relativity does not prefer a direction of time, though if there is cause and effect a time direction between two moments may be imposed. Chronology is the theory dealing with chains of moments, where for example periodic events on a given point on earth such as the sunrise, the return of the full moon, or the return of summer help to order time in calendars. The concept of chronological order is unequivocal if the location of the observer does not change. Different locations involve synchronization between calendars at different points as the sun does not rise simultaneously at all points on earth.

As we will see, there are as well hypotheses on spacetime. It includes a continuous extension in four dimensions where one coordinate is used to describe time and three coordinates are used to describe space. One point in spacetime is an event, but following the concept of Herbert Minkowski (1909) a non-Pythagorean metrics applies for the determination of the "distance" or spacetime interval between two events.

Having established the above hypotheses on space and time it is possible to elaborate from rest to motion. Motion as the antonym of rest connects with a kinematic theory points in space *and* time. Yet, rest and uniform motion should be described with the same theory. This claim will be motivated by one hypothesis in mechanics where according to Isaac Newton, the change of motion of an inertial mass requires the action of a *force* on it. The laws of nature shall have the same description in a home office or in the cabin of a ship moving in the frame of reference of the home office and vice versa. This is an expression of the relativity principle, the first hypothesis of Galilean and Einsteinian relativity (Section 1.2.1).

In contrast to speed in Galilean theory, Einsteinian special relativity claims that speed has an upper bound. The upper bound turns out to be the speed of light in vacuum and forms the second hypothesis in special relativity (Section 1.2.2). This has counterintuitive consequences and couples space and time into spacetime. A thorough evaluation of this hypothesis opens quite a different world in which our intuition and the Galilean theory only apply to small speeds compared to the speed of light. The price of abandoning intuition is rewarded by the description of mechanics and electrodynamics within the same theory for spacetime. After the introduction of the hypotheses of electrodynamics, like charge and electric fields, we will understand with the Maxwell equations the propagation of light in empty space with the same theory on space and time that describes the motion of the sun, as that of a bicycle.

1.2.1 The principle of relativity

The universality of the laws of nature was not assumed in antiquity. The world was divided into heaven and earth. Accordingly, the physics of Aristotle does not demand the same laws for the sub- and the supralunar world. For example, a stone falling to the earth did not have to follow the same principle of motion as the moon orbiting the earth. The relativity principle includes the hypothesis that nature obeys everywhere to the same universal laws. It is common to the Galilean and Einsteinian description of space and time and states that the laws can be described in any reference frame that moves with constant velocity. It implies a description of motion and rest and can be validated with an operational concept of measurements of time spans and distances everywhere in space and at every time. This can be considered as the starting point of modern physics.

1.2.2 The speed of light

Classical Galilean and relativistic Einsteinian physics are different if it comes to the discussion of all possible speeds between different reference frames. In the classical picture, every speed may be assumed, while Einsteinian physics postulates a maximum speed that is bound to the speed of light in the vacuum.

It is not surprising that the idea of a limit speed came relatively late into play. And, if there were such a limit, it was not clear to be the speed of light c. Today we know from measurements that c is 299792458 m/s or within 1‰ $3 \cdot 10^8$ m/s. This is six orders of magnitude faster than the speed of sound, which is close to any velocity reachable in mechanical devices.

Galileo Galilei proposed an experiment to measure c by covering and uncovering lanterns at different distances from each other and comparing the corresponding time lags. The experiment was performed by the *Accademia del Cimento, Firenze*, without success (Cohen, 1944). Astronomical observations of eclipses of moons of Jupiter and the timing with respect to the position of Earth (Römer, 1676) gave a first value of c.

Later, seasonal deviations of the observation angles of fix-stars during an annual cycle were interpreted as the aberration of the starlight by Bradley (1727), which reflects the ratio between the velocity of the Earth in its orbit around the sun and the speed of light. The first successful laboratory experiments to measure c were performed by Hippolyte Fizeau in 1849 and Léon Foucault in 1851 to an accuracy of better than 5%. They used one rotating and one non-rotating mirror and determined the speed of light from the distance between these mirrors, the number of revolutions per time interval of the rotating mirror and the deflection angle between the first and the second reflection on the rotating mirror. In 1972, c was determined by frequency and wavelength measurements to an accuracy of $1{:}10^9$ (Evenson et al., 1972). The accuracy of these high-precision experiments was limited by the definition of the measure of length, the meter, which eventually prompted the optical definition of the meter in favor of comparing it to the prototype standard rod in Paris. The availability of movable atomic clocks with an accuracy of 1 in 10^{14} allowed to measure time dilation, i.e. different flight times measured with a clock at rest and a clock that was on a flight around the world (Hafele and Keating, 1972a,b). Notably, the dilation effect of general relativity, i.e. of different gravitational potentials that the two clocks experienced during the flight was larger than that due to special relativity. Today these high-precision physics allow the operation of the global positioning system (GPS), which is based on the measurement of positions and times, and which reaches an accuracy down to centimeters.

1.2.3 Lorentz transformations

The Lorentz transformations are the relativistic counterpart of the Galilei transformations. Both transformations are linear and describe on how the coordinates (t, x, y, z) and (t', x', y', z') of an event map in two different frames of reference that move with constant velocity relative to each other. Like the Galilei transformations, the Lorentz transformations are derived from hypotheses on space and time. For the case of the Lorentz transformations, the time and the space coordinates are coupled, and an upper bound for relative motion emerges. For the limit of small velocities relative to the upper bound, the Lorentz transformations correspond to the Galilei transformations.

In an uncommon approach to the theory of special relativity and the Lorentz transformations that is discussed in detail in Section 2.2.2 and 2.3.2, we elaborate on the hypothesis system:

1. The two celestial spheres mapped at the moment of the coincidence by the two observers are representations of the same physical situation.
2. The celestial spheres mapped at the moment of coincidence by the two observers are conformal.

This system of hypotheses refers to local observations, i.e. to the detection of directions (angles) and frequencies of light arriving at the position of the coincidence of two local observers that may differ by their relative velocity (Komar, 1965; Blatter and Greber, 1988). It involves two concepts: aberration and Doppler shift. Notably, aberration and Doppler shift would not exist for light if the speed of light were infinite. The first hypothesis corresponds to the principle of relativity and accounts for the equivalence of the observers. It implies a reciprocity of the coincidence and the parameters defining it: the relative motion of the coinciding observers is the same except for the directions, which are opposite to each other. The second hypothesis restricts the mapping from one to the other celestial sphere to linear fractional functions as it is the stereographic projection.

In the approach based on symmetries the hypotheses
1. Spacetime is homogeneous,
2. Space is isotropic,
3. The principle of relativity is valid

are used. This system of hypotheses seems to be particularly obvious and appealing. The derivation of the Lorentz transformation from this system, without the hypothesis of the invariance of the velocity of light, yields a quantity with the dimension of a velocity that must be invariant, without claiming a value for this velocity. Empirically the value for the invariant velocity turns out to be the speed of light where the prediction of its value is an unsolved task of physics.

Various authors proposed variants of hypothesis systems for special relativity and the Lorentz transformations and a small collection is presented in the following.

The first one by Einstein (1905a),
1. Spacetime is homogeneous,
2. reciprocity of the relative velocity v,
3. invariance of the velocity of light,
4. length contraction depends only on $|v|$.

The invariance of the velocity of light is not necessary, and Ignatowsky (1910) and Frank and Rothe (1911) proposed a reduced system,
1. The transformation is linear,
2. reciprocity of the relative velocity v,
3. length contraction depends only on $|v|$,

and Bramson (1968) proposed a variant of the above systems,
1. Space is homogeneous,
2. space is isotropic,
3. invariance of the velocity of light,
4. the transformation is once differentiable,
5. the transformation is symmetric with respect to space.

A system of hypotheses proposed by Zeeman (1964) takes a different approach with statements on the metric of spacetime, causality, and the dimension of space,

1. Metric is given by a characteristic quadratic form $(ct)^2 - x^2 - y^2 - z^2$,
2. causality is conserved (partial order relation),
3. the dimension of spacetime is larger than 2.

1.3 Radical relativity

The special theory of relativity is often heuristically explained by using pictures originating in classical physics. This may work well as long as the explanations are restricted to qualitative descriptions of phenomena, if the observational setup is specified precisely and if the terminology is uniquely defined. Hermann Minkowski contributed significantly to the present day formulation of the special theory of relativity. He assigned a "radical tendency" to his view on space and time (Minkowski, 1909), as it involved a change in calling the place where physical reality is described from there on to be the spacetime.

The attribute "radical" also entered in the title of this book "Radical Relativity" as it is motivated by our intention to develop the special theory of relativity from hypotheses on space and time, geometry, and chronology, as the foundation of physics. The word "radical" is used in its original sense from the Latin "radix", meaning "root". We start from the core hypothesis of special relativity postulating a limit velocity, which steps beyond classical mechanics. As a consequence, the language for describing special relativity should be radically different from the language describing classical mechanics. In principle, different words should be used for the physical terms, so that these words do not remind one of the corresponding physical terms in classical physics.

Of course, this does not mean that we will abstain from the use of the notions of space and time, also because they can be measured. As we will see, the fact that time passes differently in frames of reference moving relative to each other implies for example different possible descriptions of speeds, except the speed of light, depending in which frame the distance and the elapsed time to cross it, is measured. It is therefore essential to carefully outline the operational measurement procedures for the determination of given quantities like the speed that may have no classical analog. This leads e.g. to the label "twin-paradox" for the different aging of two twins that move relative to each other. The label "paradox", i.e. to claim that this is not consistent with the theory (doctrine), is not correct and therefore misleading. It exemplifies the importance of using precise notions and of understanding their implications. As a further example of confusion upon mixing hypotheses, we mention "ether", the concept that relies on the assumption that light needs, like any mechanical wave, a medium such as it is air for sound that enables the propagation as a wave. Empty spacetime on its own is this "medium" in which light propagates, as it is implicit in the Maxwell equations.

1.3.1 Local and non-local observers

The word *observer* produced some of the pitfalls in special relativity. A local observer logs signals that arrive at his location. An observer in the Einsteinian sense is a set of rods and synchronized clocks evenly distributed in space and is a non-local observer. In the book of George Gamow, *Mister Tompkins in Wonderland* first published in 1939, Tompkins, a local observer, is riding a bicycle in a relativistic world showing the houses in the street contracted in the direction of his motion (Figure 1.1), as erroneously suggested to be *seen* by the cyclist. This is the perception of an Einsteinian observer, but not of Mister Tompkins, who is viewing the street from one place in space by collecting the information brought to him by light from all directions, thus he sees aberration (Figure 1.2). Clearly, this is the view of a different type of observer than the Einsteinian observer and the two views have to be distinguished. The Einsteinian observer registers length contraction and time dilation, the individual observer sees *aberration* and *Doppler shift* (Sears, 1966). Unfortunately, Einstein used the word *Beobachter* (observer) in both connections (Einstein, 1905a). To be on the safe side and not to mix the two concepts of observers, different words should be used. In this book, we use the word *observer* exclusively for the individual or local observer collecting information from the light cone at one point in space. The *Einsteinian observer* is denoted as *frame of reference.*

Fig. 1.1: Mr. Tompkins is riding a bicycle at relativistic speeds. The streets, houses, and people are depicted as Lorentz contracted as supposed to be seen by Mr. Tompkins. From Gamow (1965).

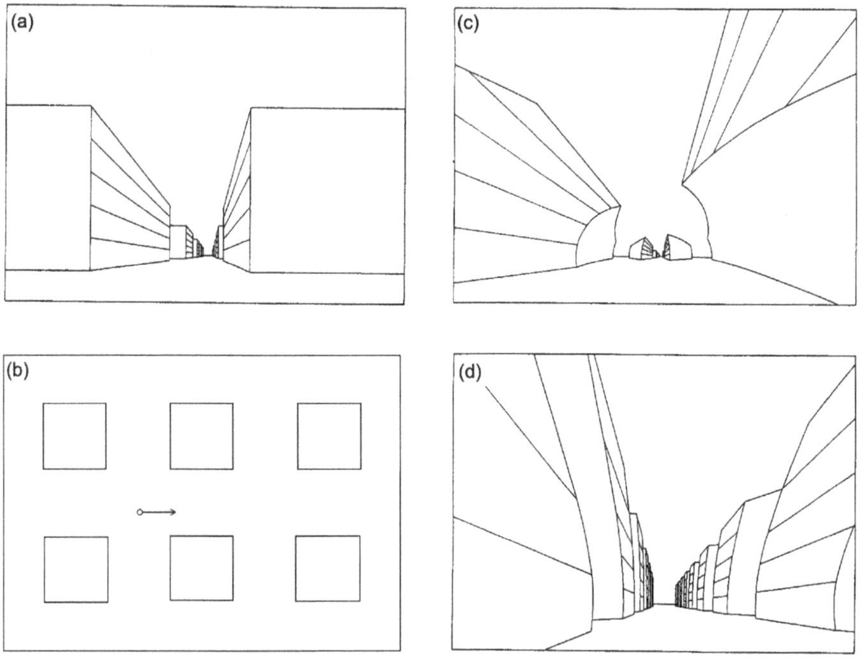

Fig. 1.2: Street canyon as seen by an observer traveling with a velocity of 90% of the speed of light in Mr. Tompkins's Relativity land. (a) View of a standing observer. (b) Top view of the street canyon and the observer (circle) moving along the arrow. (c) View of the observer in traveling direction. (d) View of the observer backward. Adopted from Blatter and Greber (1988).

1.3.2 Observables in space and time

Observables, such as angles, distances or time intervals, are quantities that can be measured with a given procedure. For example, a distance to a mirror may be inferred from the runtime of a light pulse, if the speed of light is known. In this Section we will briefly discuss three aspects of observability. First, the implications for a local observer in a Michelson-Morely experiment. Second an Einsteinian observer, where distinct clocks have to be synchronized. And finally two Einsteinian observers that move relative to each other, and who observe detuning of their clocks. i.e. time dilation.

The Michelson-Morley interference experiment (Michelson and Morley, 1887) is often referred to as the crucial ether drift experiment. It is found that the runtime of light pulses does not depend on whether they propagate along the motion of the Earth around the sun, or perpendicular to it. In the context of classical mechanics, this disproves the existence of a light-bearing ether. When interpreted in the context of special relativity this is not true anymore. An absolute space with a Lorentz invariant ether would also not change the result of the experiment. Every frame of reference with Lorentz invariance depending on the speed relative to the absolute system is also Lorentz invariant with every other reference frame, and eventually cannot be distinguished from an absolute frame (see Section 4.1), thus “the ether just fades away” (Mirabelli, 1985), and is not observable.

An Einsteinian observer or frame of reference needs to be synchronized. To explain the synchronization procedure we consider two spatially separated clocks and clockmasters A and B. Clockmaster A sends a light signal to B, B reflects it with a mirror and A registers the echo after a delay $\Delta\tau_{ABA}$. From the spatial distance between A and B, Δx_{AB}, and knowing the speed of light, A may synchronize clock B with her clock. This procedure involves the two-way velocity of light, i.e. the average of the one-way velocity between A and B and that back from B to A. Therefore, this procedure does not allow the determination of the one-way velocity of light, which is not observable with this synchronization procedure. This is still true, if B synchronizes his clock with that of A, since in both echo delays $\Delta\tau_{BAB}$ and $\Delta\tau_{ABA}$ the same two-way velocity of light is involved, and the one-way velocities can not be observed. It may, however, be postulated from the assumption of the isotropy of space or the result of the Michelson-Morely experiment.

With the same arguments, we can question the observability of time dilation and space contraction, which both depend on the convention for synchronization (see Section 4.5). If two clocks A and C move relative to each other it turns out that A has to assume a slower clock rate for C and vice versa. This also predicts space contraction, where the distance between two points depends on the reference frame in which they are observed. Importantly, if A should find a different result than C this would violate the relativity principle. This symmetry was for decades a reason for the pitfall of the twin paradox (see Section 4.6), where the “asymmetry” between the twins was not realized. As a matter of fact more than two inertial frames are needed to explain the story of the twins that are not accelerated the same way.

1.4 More Hypotheses

In the above Sections, the pivotal importance of hypotheses involving space as the “physical playground” of geometry and spacetime as that of kinematics was motivated. If we want to describe physical objects that are driven in space and time, we need to introduce hypotheses on the concepts of *mass*, *charge*, and *spin*. For the interaction between such objects in spacetime we rely on the concept of *forces* and *energy*, which is done in the Sections on mass, charge, and spin. In a further Section, we discuss the wave-particle duality as a hypothesis of quantum mechanics.

1.4.1 Mass

While it turns out that the physical description of light in the vacuum does not involve a mass (see Section 3.3.1), mass is required for the observation of light. This highlights the paramount importance of the understanding of light-matter interaction. In this context, a concept of force has to be developed, where the change of velocity of a mass and the forces between masses has to be distinguished.

Inertial mass

In mechanics we learn that moving objects are "charged" with the quality of an *inertial mass*, for which additional hypotheses are required. This is reflected in the classical *Newtonian Laws*:

1. First law: every body remains in a state of constant velocity unless acted upon by an external force.
2. Second law: a body of mass m subject to a net force F undergoes an acceleration a that has the same direction as the force and a magnitude that is directly proportional to the force and inversely proportional to the mass,

$$F = m\,a\,. \tag{1.1}$$

3. Third law: the mutual forces of action and reaction between two bodies are equal, opposite, and collinear.

The first law is the *principle of inertia*. It is uncritical in its interpretation of the classical and the special relativistic hypotheses on space and time. The inertia inherent to mass is also found in the often used term *inertial frame* in which a mass persists as long as no force acts on it. Often it is used synonymous to rest frame. The first law can be confirmed by measurements in the realm of kinematics alone, although it is difficult to make experiments with truly force-free motion. Nevertheless the principle of inertia can be confirmed by daily life experience if one travels in a train or an aircraft, where the main uniform motion can not be felt without looking outside. Only the high-frequency perturbations by shaking due to uneven rails or turbulence can be experienced directly as a deviation from uniform linear motion.

The second Law is more difficult to interpret since in the relation $F = m\,a$, three quantities are related to each other, of which only the acceleration a is defined by kinematic theory. The inertial mass m should be defined by a hypothesis, and the force itself also requires a definition. Thus far, the second Law establishes a circular definition of two quantities with one equation. At this stage of the construction of systems of hypotheses, no resolution of this circularity is given, but the circularity itself does not necessarily imply a contradiction. The second law stems from classical physics and is invariant under Galilean transformations, it can be considered as a limit of the special relativistic second Law for vanishing velocities. Importantly, as shown in Section 3.2, the relativistic

second Law can be obtained from the classical counterpart and the Lorentz transformation, and no additional hypothesis like that of a relativistic mass is necessary. Notably, the case for $m = 0$, which applies to light imposes via Eq. (1.1) that no force may be applied on light, as long as it propagates in vacuum.
The third law (actio=reactio) is a consequence of the principle of relativity (Section 1.2.1). It imposes the cancellation of action and reaction forces and is the basis of energy conservation.

Gravitational mass

Isaac Newton (1686) successfully explained the motion of planets around the sun with the assumption of *gravitation*, as a two-body force acting between the masses of the bodies which decreases with the inverse square of their distance. This signifies an additional property of matter. With the equivalence principle (see Section 5.1) gravitational mass is set equivalent to inertial mass.

Curved spacetime: besides the inertial mass from Newton's second law and the electric charge from the Coulomb law, massive objects have *gravitating mass*. This gravitating mass introduces a novel quality to spacetime. While inertial mass limits for a given force the acceleration, gravitational mass acts on spacetime itself. This has significant consequences in both the physical phenomena and in their mathematical formulation.

Strictly, *active gravitating mass* and *passive inertial mass* on which the gravitation acts should be distinguished. Newton's law of gravitation describes the mutual attactive force F_G that two (point-like) masses m_1 and m_2 exert on each other over a mutual distance r as

$$F_G = G \frac{m_1 m_2}{r^2}, \tag{1.2}$$

with the coupling (gravitational) constant G, as a two-body interaction, where the passive (inertial) m_2 or m_1 and active (gravitational) m_1 or m_2 form of mass are equivalent. This is the content of the equivalence principle of Einstein (1916) (Section 5.1).

Newton's law of gravitation describes a force that acts instantly over any distance. It is thus not consistent with special relativity, where spacetime is such that nothing can move faster than the speed of light. Similar to the Coulomb force between electric charges, Newton's law of gravitation can at best be valid for static mass distributions. Yet, this is not possible since gravitating masses have to move to stay separated. The gravitational field is not a force field since different bodies with different masses feel different forces. With Newton's second law, the acceleration is the characteristic quantity for a given point in space. Therefore, gravitation is best described as an acceleration field.

The weightlessness inside a satellite orbiting Earth indicates that the inertial mass is equal to the gravitating mass, and this holds for any body, independent of its chemical or otherwise structural composition. An astronaut can not distinguish by physical experiments whether he sits in a spacecraft accelerating due to the thrust of the rocket engines or whether he stands on a planet with a gravitational field causing the same acceleration, provided his laboratory is small enough such that effects of inhomogeneities in the field can not be detected. This is the so-called *principle of equivalence* which stands at the foundation of the theory of gravitation and the general theory of relativity (see Chapter 5).

The weightlessness in the satellite is only locally defined. A second satellite orbiting nearby will always move slightly differently. This implies that neighboring points in an accelerating field diverge with time. In Section 5.2 we will discuss the facts suggesting that a comprehensive theory of the gravitation must be a theory of a curved spacetime.

Singularities: mass may not only curve spacetime but one can think of singularities where the curvature diverges. It is suggested that this happens inside a black hole. From Eq. (1.2) the escape velocity v_e as a function of the radius r may be determined. In particular one finds for $v_e = c$ the so-called Schwarzschild radius as the critical radius of a sphere containing a mass m_1 from where the escape velocity is the speed of light. If the radius were smaller, though with the same mass embedded, no light would escape. This corresponds to a horizon where no communication between the inside and the outside is possible (see Sections 5.3.2 and 5.3.3) and coined the term "black hole".

1.4.2 Charge

Classical electrostatics is introduced by the Coulomb law describing the force between two point-like bodies 1 and 2 with charges q_1 and q_2

$$F_C = -\frac{1}{4\pi\epsilon_0}\frac{q_1 q_2}{r^2}\,, \tag{1.3}$$

where r is the distance between the two charges, and $1/4\pi\epsilon_0$ the Coulomb constant that defines the strength of the force. ϵ_0 is called the vacuum permittivity. The Coulomb force obeys Newton's third law where the force of q_1 on q_2 is equal and opposite to that of q_2 on q_1. The Coulomb force may be either attractive or repulsive. The sign in Eq. (1.3) indicates repulsion for charges with the same sign, and attraction for charges with the opposite sign.
The force on body 2 is mediated by an electric field produced by q_1

$$\mathbf{F}_C = q_2\,\mathbf{E}_1(\mathbf{r})\,, \tag{1.4}$$

with $\mathbf{r} = \mathbf{r}_2 - \mathbf{r}_1$.

As soon as the charges are moving relative to each other, magnetic fields and the so-called Lorentz force emerge. Electric and magnetic fields can be understood in the framework of the Maxwell equations, which are the foundations of electrodynamics and which describe the interplay between charges. There is no perfect "twin" symmetry between the electric and the magnetic fields since there appear, in contrast to electric charges, no magnetic charges as sources of electromagnetic fields in nature. More importantly the Maxwell equations have solutions for the vacuum, which is electromagnetic radiation propagating with the speed of light c (see Section 3.3).

The Maxwell equations and the exploration of their solutions are an immense success of a single physical theory. Therefore, all hypotheses related to the Maxwell equations should be valid and it is, likely the strongest evidence for the special theory of relativity.

1.4.3 Spin

Spin is a property of the constituents of matter besides mass and charge. Although it can be measured like the magnetic moment of a charge current in a loop, it is not classical as it is the mass or the charge. In quantum mechanics, it has deep implications like the Pauli principle. With the Pauli principle and the solutions of the Schrödinger equation, the periodic table of the elements can be understood. The spin was first observed by Gerlach and Stern (1922). They found that an atomic beam of silver atoms in an inhomogeneous magnetic field was split into two components. This ambiguity is resolved if the electron may take two orientations, i.e. *up* or *down*, and could be explained after the introduction of the spin quantum number, which is +1/2 or -1/2 for an electron.

1.4.4 Wave-Particle duality

The wave nature of particles, and the particle nature of waves form the base of quantum mechanics.

In quantum mechanics, the hypothesis of the wave function is a central concept. The square of the wavefunction describes the probability density to find a particle at a given point in space and time. The concept of the observer acquires a new meaning in the sense that every observation involves the change of the wavefunction. Still, the observers as discussed in Section 1.3.1 are compatible with this implication.

Photons

The concept of light as an ensemble of massless particles, rather than waves, is neither a consequence of special relativity nor in contradiction with it. The photon hypothesis foots on the second 1905 paper of Einstein (1905b), where he explained the light frequency threshold in the photoionization of matter. The frequency of light, not the intensity decides on whether an electron can be emitted from an illuminated body. The photoelectric effect can be understood if the light is composed of particles or quanta with an energy E_p proportional to its frequency ν

$$E_p = h\,\nu\,, \tag{1.5}$$

where the proportionality constant h is the Planck constant. If this photon energy E_p exceeds the ionization threshold, photo-ionization may be observed by the detection of electrons or ions.
Arthur H. Compton (1923) showed experimentally that photons also carry momentum along their direction of propagation,

$$p_p = \frac{h}{\lambda} = \frac{E_p}{c} = \frac{h}{c}\nu\,, \tag{1.6}$$

with λ being the wavelength of light.

Finally, it was shown by Raman and Bhagavantam (1931) from the interpretation of optical spectra and by Richard A. Beth who directly measured the angular momentum transfer of circularly polarized light (Beth, 1935) that photons carry angular momentum along their direction of propagation,

$$L_p = \pm\frac{h}{2\pi}\,. \tag{1.7}$$

Atoms

For objects like atoms that consist in electrons or protons Louis de Broglie's hypothesis that massive particles behave as waves with a wavelength λ that is inversely proportional to their linear momentum and a formula analogous to that of the massless photons in Eq. (1.6) applies,

$$\lambda = \frac{h}{p}\,, \tag{1.8}$$

where $p = mv$ is the momentum of the particle. With this, atoms can be understood as containing electron standing waves. The apparent stability of atoms furthermore required Bohr's postulate that the electrons must not lose energy via radiation as is expected for accelerated charges in classical electrodynamics. The length scale of atoms is the Bohr radius $a_0 = 52.9$ pm that depends on the Planck constant, electron mass m_e, electron charge q_e, and the Coulomb constant $1/4\pi\epsilon_0$

$$a_0 = \frac{\epsilon_0 h^2}{\pi m_e q_e^2}\,, \tag{1.9}$$

The hydrogen atom is well described with the solutions of the Schrödinger equation, which is non-relativistic in the sense that time appears as an independent variable. The relativistic extension of the Schrödinger equation is the Dirac equation, which led to the prediction of antimatter. To analyze whether relativistic effects are important in atoms we should compare the *velocity* of an electron on an orbit with radius a_0 with the speed of light c. The electron velocity on this orbit, the Bohr velocity v_B, is

$$v_B = \frac{\hbar}{m_e a_0} \equiv \alpha \cdot c\,, \tag{1.10}$$

where $\alpha = 1/137$ is the fine structure constant. The fine structure constant is an essential number as it is a measure of the strength of the interaction of light with matter. It is relatively weak and profoundly affects the appearance of the world. Along this line, we expect no strong relativistic effects in the hydrogen atom, though, the velocity in the lowest Bohr orbit increases proportionally to the square of the number of protons in the nucleus, Z, and thus relativistic effects like the spin-orbit interaction become important for heavier atoms.

2 Theory

A theory of space and time is required for any description of location and motion of physical objects. The interactions between objects with mass and charge depend not only on the relative location (distance), but as well on the relative motion between them. Eventually, this implies that space and time are connected via the speed of light. All this physics is considered by the special theory of relativity. The quantitative descriptor of motion is “velocity”, which describes a change in location Δs in a given time interval Δt. We will use the concept of coordinate velocity $\beta = v/c$, where v is $\Delta s/\Delta t$ as measured in a given frame of reference and where c is the finite speed of light that may not be surpassed.

With this the special theory of relativity i.e. a theory of spacetime is developed in this Chapter. We start with the consequences of the observations at one point, and then expand the concept to the 4-dimensional Minkowski space that arranges events in spacetime.

2.1 Coordinate systems and Coordinates

A point in space is an entity that is completely described by its location. The location or position of the point is unequivocally characterized by n numbers, where n is the dimension of the space. The numbers, also called coordinates, depend on the corresponding coordinate system. Cartesian coordinate systems are most common. In 3 dimensions (n=3) the coordinates are labeled x, y and z. They indicate the distance from the origin along the three corresponding axes $\boldsymbol{x}, \boldsymbol{y}, \boldsymbol{z}$. The axes point along unit vectors $\boldsymbol{e}_x$, $\boldsymbol{e}_y$ and $\boldsymbol{e}_z$ and are orthogonal. The sequence of axes $\boldsymbol{x}$, $\boldsymbol{y}$, $\boldsymbol{z}$ is by convention a right handed system $\boldsymbol{e}_x$ (right thumb), $\boldsymbol{e}_y$ (right index finger), $\boldsymbol{e}_z$ (right middle finger). The origin O at $(x_O, y_O, z_O) = (0, 0, 0)$ should have no specific (absolute) role if space is invariant under translations, but often it is used to describe the position of a local observer. The connection between the origin O and a point P is the location vector $\boldsymbol{r}_\mathrm{P}$ of the point with respect to O, $\boldsymbol{r}_\mathrm{P} = x_\mathrm{P}\boldsymbol{e}_x + y_\mathrm{P}\boldsymbol{e}_y + z_\mathrm{P}\boldsymbol{e}_z$. The distance to the origin is measured with the Pythagorean formula

$$d_\mathrm{OP} = \sqrt{x_\mathrm{P}^2 + y_\mathrm{P}^2 + z_\mathrm{P}^2}\,. \tag{2.1}$$

For three distinct points A, B, C in a space with $n > 1$ we find a triangle, if A, B and C do not lie on a line. If A is chosen to lie at the origin and to be the vertex, the cosine of the angle α between $\boldsymbol{r}_\mathrm{AB}$ and $\boldsymbol{r}_\mathrm{AC}$ is given by the scalar product of $\boldsymbol{r}_\mathrm{AB}$ and $\boldsymbol{r}_\mathrm{AC}$ and the distances d_AB and d_AC,

$$\cos\alpha = \boldsymbol{r}_\mathrm{AB}\cdot\boldsymbol{r}_\mathrm{AC}/(d_\mathrm{AB}\,d_\mathrm{AC})\,. \tag{2.2}$$

For $n = 3$ it is often convenient to use spherical coordinates, where the coordinates of a point consist in the distance $d_\mathrm{OP} = r_\mathrm{P}$, and two angles θ_P and φ_P. The polar angle θ_P is

https://doi.org/10.1515/9783111503592-003

measured from the north-pole (z-axis), and the azimuthal angle φ_P in the x-y plane from the x-axis, where the sense of rotation in a right handed system is counter-clockwise. Figure 2.1 shows the Cartesian and spherical coordinates (x_P, y_P, z_P) and $(r_P, \theta_P, \varphi_P)$ of a point P with respect to the origin O.

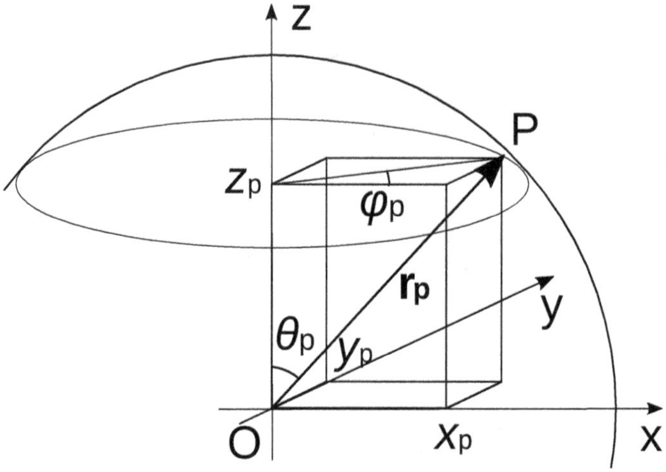

Fig. 2.1: Cartesian and spherical coordinate systems centerd at the origin O. The axes x, y, z are orthogonal. The coordinates (x_P, y_P, z_P) and $(r_P, \theta_P, \varphi_P)$ of the point P are indicated. The polar angle θ is measured from the north-pole (z-axis), and the azimuthal angle φ in a x-y plane from the x-axis. The sense of rotation or sequence of axes x, y, z is by convention a right handed system.

The transformation between Cartesian coordinates (x_P, y_P, z_P) and spherical coordinates $(r_P, \theta_P, \varphi_P)$ coordinates is

$$x_P = r_P \sin\theta_P \cos\varphi_P , \quad y_P = r_P \sin\theta_P \sin\varphi_P , \quad z_P = r_P \cos\theta_P . \tag{2.3}$$

The use of spherical coordinates is convenient if it is about comparison of two coordinate systems that move relative to each other. Often the direction of motion is chosen to be the x-axis, or the z-axis. As aberration i.e. the angles or directions of appearances are conveniently measured relative to the direction of motion, we chose the z-axis as the direction of motion and the polar angles θ and $\pi - \theta$ as the angles of appearance. The directions θ and $\pi - \theta$ are opposite as they are registered in coordinate systems moving relative to each other.

2.2 Local observers: aberration and Doppler shift

Aberration and Doppler shift are phenomena encountered by local observers, who move a detector relative to a source that emits signals in the form of "particles" or "waves" with a certain velocity and frequency. In the case of aberration, the detector consists in a goniometer that measures angles between the direction of motion and the arrival of the signal. For the Doppler effect, a clock measures frequencies or arrival rates of signals. As we will see, these at first sight classical effects open a door to special relativity.

2.2.1 The everyday phenomena

Aberration and Doppler shift are known from everyday life. They are related to the relative motion of a source (sender) with respect to an observer (receiver). The source may for example emit rain, sound, or electromagnetic radiation with a certain velocity and frequency. In the case of rain, we may experience aberration while running. If the drops fall vertically, as soon as one starts to move, the raindrops appear to fall from ahead and an umbrella is tilted forward for better protection (see Figure 2.2). The tilt angle of the umbrella depends on the ratio of the speed of the falling raindrops and the relative speed of the umbrella. The Doppler effect is reflected in the change of the arrival rate of raindrops for different velocities.

Aberration (and Doppler shift) can also be observed with a passing airplane. We associate the location of the airplane with the direction where we can see it. The optical image of the airplane appears always ahead of the direction, from where the sound of the airplane is coming because the speed of light is larger than that of the sound. Now, imagine that the airplane is not visible, either you close your eyes or fog obscures the sky. In this case, the location of the airplane can only be determined by the direction and the frequency of the noise of the arriving sound, and the ratio between the speed of sound and the speed of light.

Aberration allows us to determine the speed of the arriving particles or waves. For light, this was first done by James Bradley (1727) when he explained seasonal variations in the angles of observation of fixed stars with the orbital velocity of the earth around the sun and a finite speed of light. Since the speed of light is 10^4 times larger than the orbital velocity of the earth, Bradley was unable to distinguish whether the speed of light depends on the detector angle. The corresponding determination of the speed of light via the seasonal change of the apparent color of fix stars was performed much later by Christian Doppler (1842).

These classical experiments are the starting point for our derivation of special relativity from aberration and Doppler shift. i.e. the hypothesis that the images of the celestial spheres of two coinciding observers are conformal, i.e. maintain the angles between fix stars.

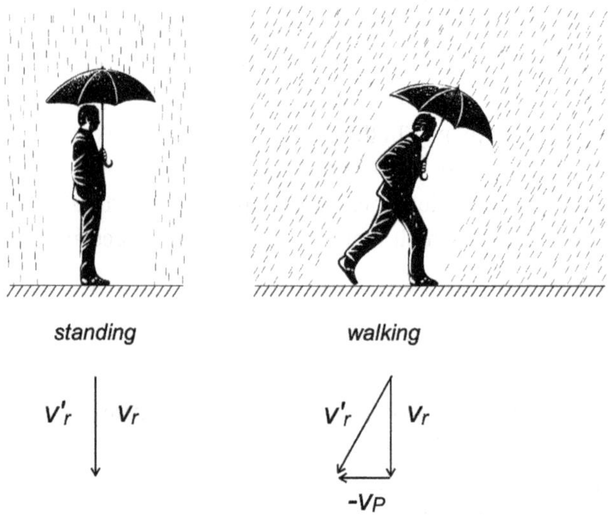

Fig. 2.2: Non-relativistic aberration and Doppler shift as experienced in walking with different velocities relative to falling rain adapted by Véron and Greber (2024) from Hoffmann (1983). The scenario on the left (standing) shows the orientation of the umbrella for maximum protection of a person at rest perpendicular to the falling drops. On the right (walking), the situation for a person with an umbrella moving relative to the scenario on the left is shown. The bottom panels depict the related velocities and how they are added. v_r is the velocity of the raindrops in the frame of reference of the ground, $v_{r'}$ that in the frame of references of the person with the umbrella, with (right) and without (left) velocity v_P with respect to the ground. Note, in the frame of reference of the person the ground moves with $-v_P$. The arrival angle and rate of the raindrops depend on the velocities and are described with aberration and Doppler shift, respectively.

2.2.2 An uncommon way to relativity: conformal maps of celestial spheres

The observer is considered to move through space in a spacecraft, from which he can map the celestial sphere, that is, he can locate each observed object i, e.g. a star, by two coordinates. In a spherical coordinate system this are the angles θ_i and φ_i (see Section 2.1). The distance to the object r_i may not be determined without further assumptions. For the moment, we do not consider the possibilities of the astronaut to determine his motion with respect to any visible object, and the spacecraft is considered not to rotate. Let us suppose that spacecraft O meets a second spacecraft O' at one given moment in time. Both astronauts map their celestial spheres at that moment and later exchange these maps by some means of radio transmission. This coincidence shall be considered a coincidence at one point in space and thus, at one given moment in time.

We now postulate two hypotheses concerning the two maps of the celestial spheres taken by the two observers at the moment of their coincidence (Komar, 1965):

1. The two celestial spheres mapped at the moment of the coincidence by the two observers are representations of the same physical situation.

2. The celestial spheres mapped at the moment of the coincidence by the two observers are conformal.

The first hypothesis, a form of the principle of relativity, states that the two coinciding observers collect their information from the same retarded light cone. This hypothesis is only valid if the spatial extent of the observers and the distance during the close encounter are negligibly small in comparisopn with the distances to the mapped stars. Furthermore, the time required for the mapping of the celestial spheres must be small enough such that no recordable changes may occur during the process.

The second hypothesis may be a result of accurate and careful analysis and comparison of the two maps. Firstly, the two maps allow to identify a particular direction in space given by the direction of the relative motion of the two observers at the moment of coincidence. For further considerations, we make an additional restrictive assumption (Blatter and Greber, 1988):

3. All vectorial quantities that define the coincidence are collinear.

With this direction, a cylindrical symmetry with two fixed points for the conformal mapping between the two maps is defined. As an arbitrary convention, we call the direction of motion of observer O' with respect to observer O the direction towards the *apex* A, and the reverse direction towards the *antapex*. For an easier treatment of the conformal mapping, the celestial spheres are projected stereographically onto a plane, where the antapex is chosen as the centre of the projection. Since the stereographic projection is conformal, the conformal mapping of the sphere onto itself is thus transformed into a conformal mapping of a plane onto itself. The conformal mapping of the plane reflects the cylindrical symmetry such that the image of the apex is a proper fixed point and the image of the antapex is an improper fixed point. The cylindrical symmetry further ensures that only the angle between the line of sight and the axis of symmetry changes, thus the conformal mapping of the plane must be a projection from the fixed point.

A conformal mapping in the complex plane is described by a Möbius transformation,

$$\mathfrak{w} = f(\mathfrak{z}) = \frac{a\mathfrak{z} + b}{c\mathfrak{z} + d}, \tag{2.4}$$

where the constants a, b, c and d, and the variables $\mathfrak{z}$ and $\mathfrak{w}$ are complex numbers. The complex plane can always be defined such that the proper fixed point is $\mathfrak{z} = 0$, thus $b = 0$. The improper fixed point requires that $c = 0$. Thus, Eq. (2.4) reduces to

$$\mathfrak{w} = f(\mathfrak{z}) = \frac{a}{d}\mathfrak{z} = Q\,\mathfrak{z}. \tag{2.5}$$

Since the mapping is a projection from the center $\mathfrak{z} = 0$, Q must be a positive real number. Thus, the mapping is a central expansion with the expansion factor Q. Finally, the conformal mapping of the celestial spheres can be described in terms of the angles θ

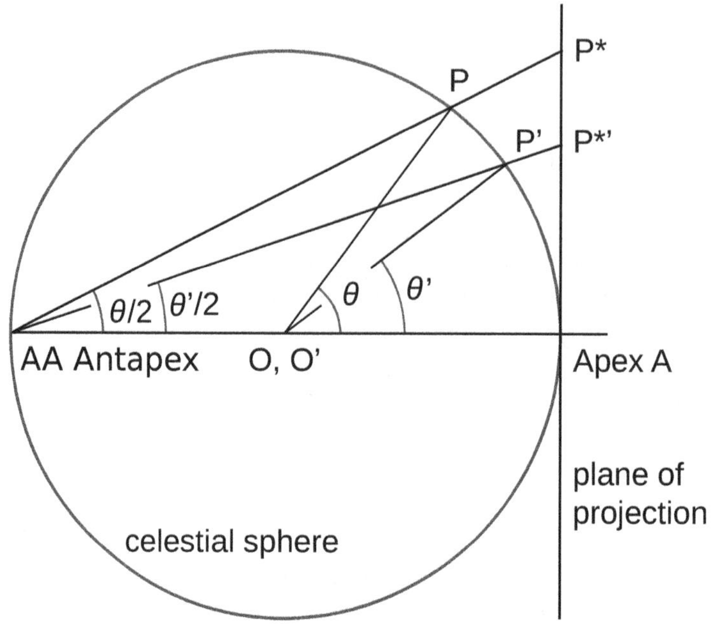

Fig. 2.3: Schematic of the stereographic projection of the celestial spheres of two coinciding observers O and O'. O' is considered to move relative to O in the direction of the apex A. The projection center is the antapex AA. The objects P and P' are seen by observers O and O' at angles θ and θ' with respect to the direction of the apex. The projections of P and P' on the plane of projection perpendicular to the direction of the apex are P* and P*', respectively.

and θ' between the lines of sight and the apex, see Figure (2.3),

$$\tan\frac{\theta}{2} = Q\tan\frac{\theta'}{2}. \tag{2.6}$$

Equation (2.6) provides the basis for the geometric representations of the aberration: the stereographic projections of the celestial spheres from the antapex AA onto the tangent plane at the apex A are related by a central stretching or expansion. The stretching factor is given by the ratios in $\overline{AP^{*\prime}} : \overline{AP^{*}}$. In Figure 2.4 a few examples of viewing angles and their mappings are plotted.

The above hypotheses compare the celestial spheres of coinciding observers at the moment of their coincidence. Coincidence is a symmetric and transitive relation, and thus, the transformations form an algebraic group. The group operation is the composition of transformations, in the special case of more observers coinciding at the same moment with relative motions in the same direction. In the case of three observers, from

$$\tan\frac{\theta_1}{2} = Q_{12}\tan\frac{\theta_2}{2} \text{ and } \tan\frac{\theta_2}{2} = Q_{23}\tan\frac{\theta_3}{2} \tag{2.7}$$

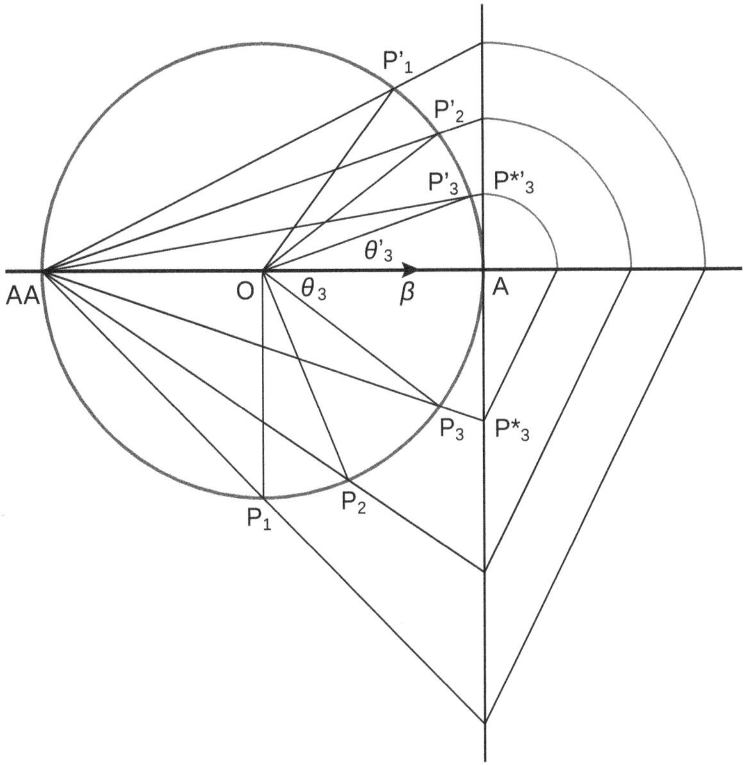

Fig. 2.4: Geometric representation of the aberration based on the central stretching of the stereographic projection from the antapex AA of the celestial spheres of two coinciding observers O and O'. On the lower (upper) hemisphere, 3 directions $\overline{OP}_i$ ($\overline{O'P'}_i$) with angles θ_i (θ_i'), are stereographically projected onto the plane of projection (tangential to apex A) where the P_i^{*}' appear centrally compressed to the P_i^*.

follows

$$\tan\frac{\theta_1}{2} = Q_{13}\tan\frac{\theta_3}{2} \text{ with } Q_{13} = Q_{12}\,Q_{23}\,. \tag{2.8}$$

The Factors Q are characteristic quantities of the magnitude of the relative motions of the observers with a multiplicative composition law. It is straightforward to define an additive quantity ρ to characterize the magnitude of the relative motion,

$$Q \equiv e^{\rho}\,. \tag{2.9}$$

The quantity ρ is called rapidity, see Section 3.1.2.

With Eq. (2.6) the system of hypotheses of Komar (1965) is mostly exploited. With the parameter Q the type of coincidence is characterized. Q can be interpreted in different physical ways not given by the hypotheses. Aberration was characterized by the determination of directions from one vantage point, only. Now, this will be extended to a concept of a geometric space in which distances are defined in addition to directions.

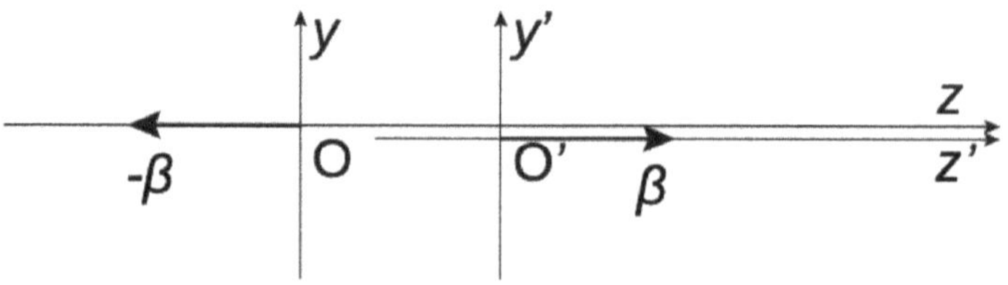

Fig. 2.5: Schematic of the coordinate systems and the relative motion of two observers O and O' at the origins of the corresponding systems. O' moves with β relative to O, while O moves with -β relative to O'.

Furthermore, the moment of the coincidence will be embedded in a time and motion will be understood in terms of the change of position in a given time interval.

Finally, we need a connection of purely mathematical statements with the physical experience. In principle, it is possible to add additional axioms and additional parameters to develop a consistent system of statements and equations, which corresponds to the mathematical statements of special relativity. We do not follow this path of reasoning but we connect to the classical experience with time and space. The inconsistency of the classical view with the axioms of Komar can be used to modify the classical theory. With this example, we demonstrate the power of heuristic reasoning, which was so important for Einstein.

A remarkable feature of the form (2.6) of the equation of aberration is that it is independent of the number of vectorial quantities characterizing aberration, as long as they are collinear. The question remains which quantities, scalar or vectorial, do determine the coincidence. It is obvious that the velocity β and acceleration α are plausible candidates,

$$Q = Q(\beta, \alpha)\,. \tag{2.10}$$

To obtain a unique parameterization of Q, which is commonly known as the Doppler factor, we define a scenario (Figure 2.5): two local observers, O and O', are situated in the origins of their respective Cartesian frames of references, the corresponding axes being parallel and oriented in the same direction. The coincidence of the two observers is given at the moment when the two origins coincide. Observer O' is considered to move along the z-axis of the frame of O with the velocity $\beta > 0$, if the direction of motion is in the direction of the positive z-axis. Reversely, observer O may be considered to move relative to observer O' in the opposite direction with velocity $-\beta$.

2.2.3 Aberration

The form of function Q cannot be determined with the above hypotheses. A possible determination may start with the equation for the classical aberration,

$$\tan\theta = \frac{\sin\theta'}{\cos\theta' - \beta}. \tag{2.11}$$

The scenario corresponds to the one shown in Figure 2.6 with the observer O' moving to the right with respect to observer O. The angles θ and θ' are measured between the positive z-axis and the incoming light beam, which in this case points from right to left (Figure 2.6). Notably in classical aberration the speed of the incoming waves or particles depends on the relative motion between source and observer. As a consequence Equation 2.11 cannot be transformed to Eq. (2.6) and thus, contradicts the hypothesis of conformal mapping of celestial spheres. We assume that the classical aberration is an approximation of the conformal aberration, and can be corrected by a factor B,

$$\tan\theta = \frac{B\sin\theta'}{\cos\theta' - \beta}, \tag{2.12}$$

which together with the tangent half angle formulae yields a unique solution for B,

$$B(\beta) = \sqrt{1-\beta^2}. \tag{2.13}$$

This makes the classical case a limit case of the relativistic counterpart for $\beta \to 0$.

Finally, the function $Q = Q(\beta)$ becomes

$$Q(\beta) = \sqrt{\frac{1+\beta}{1-\beta}}, \tag{2.14}$$

which is the correct function Q as in the established equation for the relativistic aberration,

$$\tan\frac{\theta}{2} = \sqrt{\frac{1+\beta}{1-\beta}}\tan\frac{\theta'}{2}. \tag{2.15}$$

The application of trigonometric identities to Eqs. (2.12) and (2.13) yields two more forms of the equation for the relativistic aberration of light, which may be useful depending on the given application,

$$\sin\theta' = \frac{\sin\theta}{\gamma(1-\beta\cos\theta)}, \qquad \sin\theta = \frac{\sin\theta'}{\gamma(1+\beta\cos\theta')} \tag{2.16}$$

$$\cos\theta' = \frac{\cos\theta - \beta}{1-\beta\cos\theta}, \qquad \cos\theta = \frac{\cos\theta' + \beta}{1+\beta\cos\theta'}. \tag{2.17}$$

The square root in Eq. (2.13) is an often recurring quantity in special relativity and the abbreviation

$$\gamma \equiv \frac{1}{\sqrt{1-\beta^2}} \tag{2.18}$$

is often used, were γ is called relativistic factor.

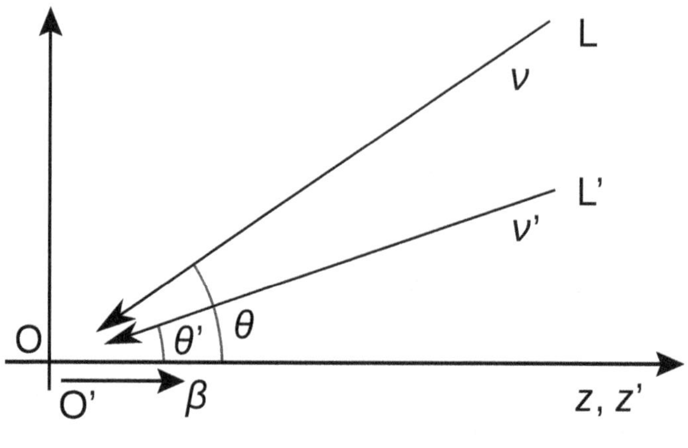

Fig. 2.6: Schematic of the Scenario for the observation of aberration and Doppler shift of two coinciding observers O and O′, where O′ is considered to move with velocity β relative to O in the direction of the z-axes. θ and θ' are the light incidence angles with respect to β, where the light sources are observed under the angles θ-π and θ'-π, respectively. L is the direction of the incoming light with frequency ν as registered by O, L' the direction of the incoming light with frequency ν' as registered by O'.

2.2.4 Doppler shift

The observation of an object in a given direction usually involves the observation of light emitted from that object. Comparison of the frequencies of the light by two coinciding observers reveals the so-called Doppler shift of light.

For the description of the Doppler shift, we define a scenario, Figure 2.6, corresponding to Figures 2.3 and 2.5 for aberration and for the relative motion of observers O and O' and their co-moving frames of reference. Observer O sees the light source with frequency ν in the direction $\overline{OP}$ at an angle θ with respect to the relative motion (apex of the motion of observer O'), and observer O' sees the light source with frequency ν' in the direction $\overline{O'P'}$ with corresponding angle θ'.

The description of aberration does not require an operational concept of time and time measurement. Observation of frequencies involves time measurement. Similarly to the derivation of the relativistic equation for aberration, the equation for the relativistic Doppler shift of electromagnetic waves can be derived with a *correction* of the classical equation for the Doppler shift, e.g. of sound waves.

Again, we consider two coinciding observers. For observer O, observer O' moves with a velocity β such that the direction of motion and the apparent direction of the incoming light include an angle θ' as observed by O'.

The frequency ν' of the light as observed by O' is Doppler shifted,

$$\nu' = \nu\,(1 - \beta\cos\theta)\,. \tag{2.19}$$

Reversely, for observer O' observer O may be considered to move in the opposite direction with the same speed, which follows from the Galilean principle of relativity, and

$$\nu = \nu' \,(1 + \beta \cos \theta')\,. \tag{2.20}$$

To solve the inconsistency between Eqs. (2.19) and (2.20) they are multiplied with a correction factor D,

$$\nu' = \nu\, D\,(1 - \beta \cos \theta)\,, \quad \nu = \nu'\, D\,(1 + \beta \cos \theta')\,. \tag{2.21}$$

Substitution of ν' in the first equation of (2.21) into the second equation of (2.21) and application of the first Eq. (2.17) yields

$$D = \gamma = \frac{1}{\sqrt{1 - \beta^2}}\,, \tag{2.22}$$

which yields the correct equation for the relativistic Doppler shift,

$$\nu' = \nu\, \gamma\,(1 - \beta \cos \theta)\,, \quad \nu = \nu'\, \gamma\,(1 + \beta \cos \theta')\,. \tag{2.23}$$

In the collinear case, $\theta = \theta' = 0$, Eqs. (2.23) reduce to

$$\nu' = \nu\, Q = \nu \sqrt{\frac{1 + \beta}{1 - \beta}}\,, \quad \nu = \nu'\, Q' = \nu' \sqrt{\frac{1 - \beta}{1 + \beta}}\,. \tag{2.24}$$

The close relation between aberration and Doppler shift can be illustrated with Eqs. (2.23). An Observer O' looking at a light source that is at rest in the frame of reference O sending light with a frequency ν sees a frequency ν' at an apparent angle θ',

$$\nu' = \frac{\nu}{\gamma\,(1 + \beta\, \cos \theta')}\,. \tag{2.25}$$

For an apparent viewing angle of $\theta' = 90°$, the Doppler shift becomes

$$\nu' = \frac{\nu}{\gamma}\,, \tag{2.26}$$

which can be called a *transverse Doppler shift*. The transverse Doppler shift is always a red shift, and even before the light source passes the transverse point at $\theta' = 90°$. Thus, the statement that an approaching light source is blue shifted and a receding light source red shifted is only true for purely radial motion, but not necessarily for apparent oblique approach or recession.

For an oblique approach of a light source, the velocity vector can be split into an apparent radial component, $\beta_r = \beta \cos \theta'$, and an apparent tangential component, $\beta_t = \beta \sin \theta'$ (Dykla, 1979). With this, Eq. (2.25) can be written as

$$\frac{\nu'}{\nu} = \frac{\sqrt{1 - \beta_r^2 - \beta_t^2}}{1 + \beta_r}\,. \tag{2.27}$$

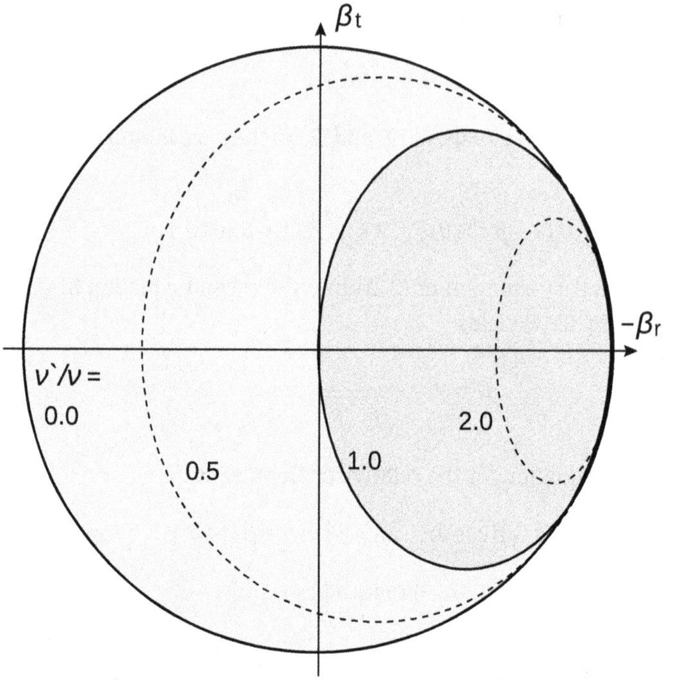

Fig. 2.7: Geometric representation of the Doppler shift on contours in a diagram of radial and tangential (vectorial) velocities β_r and β_t. The dark shaded area inside the ellipse with $\nu'/\nu = 1.0$ denotes the domain of blue shift.

This is a quadratic function of β_t and β_r, and since the graph of this function is confined to a circle with radius = 1 about the origin, the curves with constant ν'/ν must be ellipses. Rearranging Eq. (2.27) yields the normal form of an ellipse with semimajor axis a and semiminor axis b,

$$\frac{(\beta_r + \beta_0)^2}{a^2} + \frac{\beta_t^2}{b^2} = 1\,, \tag{2.28}$$

with the parameters

$$\kappa \equiv \left(\frac{\nu'}{\nu}\right)^2, \quad a = \frac{1}{1+\kappa}\,, \quad b = \frac{1}{\sqrt{1+\kappa}} \quad \text{and} \quad \beta_0 = \frac{\kappa}{1+\kappa}\,. \tag{2.29}$$

The centers of the ellipses are at ($\beta_t = 0/\beta_r = -\beta_0$). The point with $\beta_r = 1$ and $\beta_t = 0$ is singular since all ellipses meet at this point (Figure 2.7).

2.2.5 Velocity composition law

For collinear relative motion the Doppler factors Q and Q' in Eqs. (2.24) and the aberration angles Eq. (2.8) are multiplicative in a composition of transformations,

$$\sqrt{\frac{1+\beta_{12}}{1-\beta_{12}}} \cdot \sqrt{\frac{1+\beta_{23}}{1-\beta_{23}}} = \sqrt{\frac{1+\beta_{13}}{1-\beta_{13}}}, \tag{2.30}$$

which yields the collinear composition law for the quantities β,

$$\beta_{13} = \frac{\beta_{12}+\beta_{23}}{1+\beta_{12}\beta_{23}}. \tag{2.31}$$

This is the relativistic velocity composition or Einsteinan velocity addition law for three collinearly moving inertial frames 1, 2, and 3. Frame 2 moves with β_{12} in frame 1, frame 3 moves with β_{23} in frame 2, and β_{13} follows accordingly. β_{13} and Eq. (2.31) are zero for $\beta_{12} = -\beta_{23}$ as shown in Figure 2.8. For $|\beta_{12}| \leq 1$ and $|\beta_{23}| \leq 1$, $|\beta_{13}| \leq 1$ follows, which indicates that β is a bound quantity.

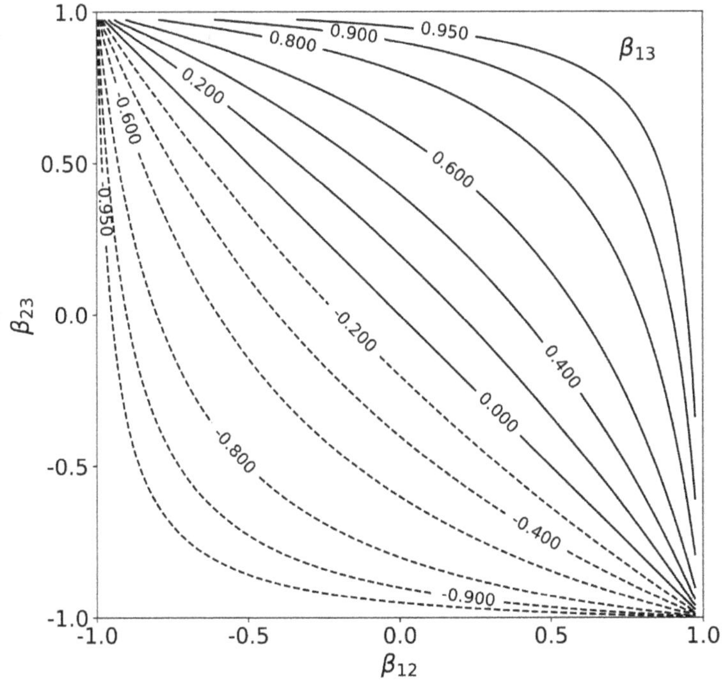

Fig. 2.8: Graphic visualisation of the velocity composition law in Eq. (2.31). The resulting velocities β_{13} are displayed as contour lines. The solutions with $-1 < \beta_{12} < 1$ and $-1 < \beta_{23} < 1$ produce a bound coordinate velocity $-1 < \beta_{13} < 1$.

For small β's ($\beta_{ij} \rightarrow 0$), Eq. (2.31) converges to the classical Gallilean velocity addition law $\beta_{13} = \beta_{12} + \beta_{23}$, while for a relativistic limit, $\beta_{12} \rightarrow 1$ or $\beta_{23} \rightarrow 1 \Rightarrow \beta_{13} \rightarrow 1$.

For the addition of non-collinear relativistic velocities the situation complicates and the phenomena of the Thomas precession (Thomas, 1927) emerges (Ungar, 1991a).

2.3 Non-local observers: Minkowski spacetime and Lorentz transformations

Section 2.2.2 describes aberration and Doppler shift of light as local observations that can be made at one point in space at a given time. Non-local, or Einsteinian observers are frames of reference or coordinate systems in which physical processes can be described for all points in space and time. Hermann Minkowski formulated the corresponding 4-dimensional spacetime concept in his seminal 1908 talk starting with:

> Die Anschauungen über Raum und Zeit, die ich Ihnen entwickeln möchte, sind auf experimentell-physikalischem Boden erwachsen. Darin liegt ihre Stärke. Ihre Tendenz ist eine radikale. Von Stund an sollen Raum für sich und Zeit für sich völlig zu Schatten herabsinken und nur noch eine Art Union der beiden soll Selbständigeit bewahren (Minkowski, 1909).
> *The views about space and time, which I would like to develop for you, have grown up on experimental-physical ground. Therein lies their strength. Their tendency is radical. From now on space for itself and time for itself shall completely sink to shadows and only a kind of union of the two shall preserve independence.*

Figure 2.9 is the first depiction of 4-dimensional spacetime (Minkowski, 1909), where light is shown as a double cone emanating into the past and the future.

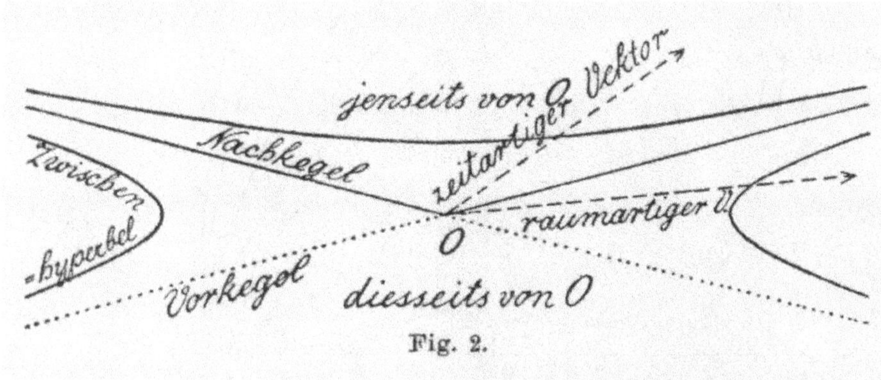

Fig. 2.9: Diagram from Minkowski (1909) connecting space (raumartiger Vektor) and time (zeitartiger Vektor). The lightcones (Vorkegel and Nachkegel) emanating from the origin O into past (diesseits) and future (jenseits) are indicated.

In this Section we will meet the Lorentz transformations as the linear mappings between different frames of reference. Here we restrict ourselves to Lorentz boosts, that treat frames of reference or coordinate systems with relative velocities along a line that is chosen to be the z-axis while the x and y axes remain invariant. Therefore the consequences of subsequent Lorentz transformations with non-collinear velocities (Ungar, 1991a) are not treated.

2.3.1 Lorentz transformations from the metrics in Minkowski space

Spacetime vectors

Spacetime is a 4-dimensional manifold, where a point is an event $\mathfrak{E}$ with the coordinates (t, x, y, z). In a different frame of reference, the same event has the coordinates (t', x', y', z'). The connection between (t, x, y, z) and (t', x', y', z') is made by a transformation which may be described as a 4×4 matrix. In contrast to Galilei transformations time and space coordinates mix. As the physical units of time (seconds) and space (meters) are distinct the off-diagonal matrix elements are not dimensionless. In order to cure this lack of elegance we multiply the time coordinate with a constant velocity that turns out to be the speed of light c. With this, we obtain transformations with dimensionless matrix elements.

In the following 4-vectors (1+3 vectors) in spacetime are denoted with boldface characters, $\boldsymbol{r}$ or $\boldsymbol{k}$. An event $\mathfrak{E}_i$ is for example fully characterized with a position vector $\boldsymbol{r}_i$. In coordinate notation, vectors are given as columns of numbers. 4-vectors are composed of 4 numbers, the top number denotes the temporal part and the three lower numbers the 3-dimensional spatial part.

Minkowski metrics: pseudo scalar product and Lorentz transformations

The step from the local observer to a concept of spacetime according to Minkowski (1909) is realized with a linear mapping between two frames of reference,

$$\boldsymbol{r}' = \boldsymbol{L} \cdot \boldsymbol{r}\,, \tag{2.32}$$

where $\boldsymbol{L}$ is the transformation matrix and $\boldsymbol{r}'$ and $\boldsymbol{r}$ are the corresponding vectors in the two frames of reference,

$$\begin{pmatrix} c\,t' \\ x' \\ y' \\ z' \end{pmatrix} = \begin{pmatrix} a_{11} & a_{12} & a_{13} & a_{14} \\ a_{21} & a_{22} & a_{23} & a_{24} \\ a_{31} & a_{32} & a_{33} & a_{34} \\ a_{41} & a_{42} & a_{43} & a_{44} \end{pmatrix} \cdot \begin{pmatrix} c\,t \\ x \\ y \\ z \end{pmatrix}. \tag{2.33}$$

In vector representation, vector algebra in spacetime is the same as in 3-space except that the concept of distance or the metrics requires a modification. Let $\boldsymbol{r}$ indicate an event, or a point in the Minkowski space. We assume a non-positive symmetric bilinear form, or a pseudo-scalar product. The term "pseudo" is chosen in order to distinguish it explicitly from the common scalar product that is always positive. The sign of the spatial part must be defined. In this book we use the convention to take it negative,

$$\boldsymbol{r}\cdot\boldsymbol{r} = \begin{pmatrix} ct \\ x \\ y \\ z \end{pmatrix} \cdot \begin{pmatrix} ct \\ x \\ y \\ z \end{pmatrix} \equiv (ct)^2 - x^2 - y^2 - z^2 \,. \tag{2.34}$$

This is shown with the assumption that the pseudo-scalar product is independent of the coordinate system, as it is the case in 3-space. However, to start, we only assume a 2-dimensional spacetime with ct and z as coordinates. With this assumption, we define the pseudo-scalar product of vector $\boldsymbol{r}$ and its transform $\boldsymbol{r}'$,

$$\boldsymbol{r}' = \begin{pmatrix} ct' \\ z' \end{pmatrix} = \begin{pmatrix} a_{11} & a_{14} \\ a_{41} & a_{44} \end{pmatrix} \cdot \begin{pmatrix} ct \\ z \end{pmatrix} = \begin{pmatrix} a_{11}\, ct + a_{14}\, z \\ a_{41}\, ct + a_{44}\, z \end{pmatrix} = \boldsymbol{r} \,. \tag{2.35}$$

This pseudo-scalar product must be independent of the coordinates,

$$\begin{aligned} & \boldsymbol{r}'\cdot\boldsymbol{r}' \\ = \; & (ct')^2 - z'^2 \\ = \; & (a_{11}\, ct + a_{14}\, z)^2 - (a_{41}\, ct + a_{44}\, z)^2 \\ = \; & (a_{11}^2 - a_{41}^2)\,(ct)^2 + 2\,(a_{11}\, a_{14} - a_{41}\, a_{44})\, ct\, z + (a_{14}^2 - a_{44}^2)\, z^2 \\ = \; & (ct)^2 - z^2 \\ = \; & \boldsymbol{r}\cdot\boldsymbol{r} \,, \end{aligned} \tag{2.36}$$

which gives 3 equations for 4 variables,

$$a_{11}\, a_{14} - a_{41}\, a_{44} = 0 \,, \tag{2.37}$$

$$a_{11}^2 - a_{41}^2 = 1 \,, \tag{2.38}$$

$$a_{14}^2 - a_{44}^2 = -1 \,. \tag{2.39}$$

Eliminating the unknowns a_{11} and a_{44} with the substitution method

$$a_{11} = \sqrt{1 + a_{41}^2} \quad \text{and} \quad a_{44} = \sqrt{1 + a_{14}^2} \,, \tag{2.40}$$

we obtain

$$a_{14}\, \sqrt{1 + a_{41}^2} = a_{41}\, \sqrt{1 + a_{14}^2} \tag{2.41}$$

which finally yields

$$a_{14} = a_{41} \tag{2.42}$$

and with Eq. (2.40)

$$a_{11} = a_{44} \,. \tag{2.43}$$

Thus, with the assumption of a pseudo-scalar product with a positive temporal part and a negative spatial part and its independence of the coordinate system, the transformation matrix is symmetric.

We obtain the same result with the full 4-dimensional spacetime, where we obtain 10 equations for 16 unknowns. If the relative motion of the systems is in the direction of the z-axis, we obtain the same result for the above matrix elements, with a solution $a_{12} = a_{13} = a_{21} = a_{23} = a_{24} = a_{31} = a_{32} = a_{34} = a_{42} = a_{43} = 0$ and $a_{22} = a_{33} = 1$.

We write a_{14} as a product of a_{11} with another parameter V, $a_{14} = a_{11}\,V$, which can be done without loss of generality. With this, Eq. (2.38) becomes

$$a_{11}^2 - a_{11}^2 V^2 = 1 \tag{2.44}$$

and finally for a_{11}

$$a_{11}^2 = \frac{1}{1 - V^2} \quad \text{or} \quad a_{11} = \pm\frac{1}{\sqrt{1 - V^2}} \tag{2.45}$$

and for a_{14}

$$a_{14}^2 = \frac{V^2}{1 - V^2} \quad \text{or} \quad a_{14} = \pm\frac{V}{\sqrt{1 - V^2}}\,. \tag{2.46}$$

Therefore, we can write the transformation with one single parameter V that must be function of the relative velocity β of the coordinate systems.

Now we assume two coordinate systems (1+1 dimensional spacetime) which move with respect to each other in the z-direction. A point on the z-axis in the dashed system with coordinate z_1' is at a point with coordinate z_1 in the unprimed system at time t_1. At time t_2, the point has traveled to a new point with coordinate z_2. The corresponding position vectors in spacetime are

$$\boldsymbol{r}_1 = \begin{pmatrix} c\,t_1 \\ z_1 \end{pmatrix} \quad \text{and} \quad \boldsymbol{r}_2 = \begin{pmatrix} c\,t_2 \\ z_2 \end{pmatrix}, \tag{2.47}$$

$$\boldsymbol{r}_1' = \begin{pmatrix} c\,t_1' \\ z_1' \end{pmatrix} \quad \text{and} \quad \boldsymbol{r}_2' = \begin{pmatrix} c\,t_2' \\ z_2' \end{pmatrix}, \tag{2.48}$$

with $z_1' = z_2'$. The difference vectors are

$$\Delta\boldsymbol{r}_1 = \begin{pmatrix} c\,(t_2 - t_1) \\ z_2 - z_1 \end{pmatrix} \quad \text{and} \quad \Delta\boldsymbol{r}_1' = \begin{pmatrix} c\,(t_2' - t_1') \\ 0 \end{pmatrix}. \tag{2.49}$$

With the transformation, Eq. (2.35), we obtain

$$c\,\Delta t' = a_{11}\,c\,\Delta t + a_{14}\,\Delta z \quad \text{and} \quad 0 = a_{41}\,c\,\Delta t + a_{44}\,\Delta z\,, \tag{2.50}$$

and therefore, with the right side of Eq. (2.50) and Eqs. (2.45) and (2.46),

$$\frac{\Delta z}{c\,\Delta t} = -\frac{a_{41}}{a_{44}} = \pm V\,. \tag{2.51}$$

With this, we can identify V as the velocity β, see e.g. Figure 2.5. We choose the times t and t' to run in the same direction in both systems, thus

$$a_{11} = a_{44} = \frac{1}{\sqrt{1-\beta^2}} \equiv \gamma\,. \tag{2.52}$$

Consequently,

$$a_{14} = a_{41} = -\beta\gamma\,, \tag{2.53}$$

and thus, the transformation, called Lorentz transformation, is

$$\boldsymbol{r}' = \begin{pmatrix} \gamma & -\beta\gamma \\ -\beta\gamma & \gamma \end{pmatrix} \cdot \boldsymbol{r}\,. \tag{2.54}$$

For this derivation of the Lorentz transformation, we used several hypotheses. The first is the assumption of a linear transformation, such as the Galilean transformation in classical physics, and the second assumption is the existence of a universal constant velocity, Eq. (2.33). The third assumption concerns the non-positive definite pseudo-scalar product Eq. (2.34).

Failure of the positive definite scalar product

The choice of the non-positive definite pseudo-scalar product for the metrics in Minkowski space requires some explanation. If we assume instead a positive definite scalar product and repeat the procedure as above, from Eq. (2.34) to Eq. (2.54), we obtain the result

$$\boldsymbol{r}' = \begin{pmatrix} \dfrac{1}{\sqrt{1+\beta^2}} & \pm\dfrac{\beta}{\sqrt{1+\beta^2}} \\ \pm\dfrac{\beta}{\sqrt{1+\beta^2}} & \dfrac{1}{\sqrt{1+\beta^2}} \end{pmatrix} \cdot \boldsymbol{r}\,. \tag{2.55}$$

With this transformation, velocities can be unlimited and c becomes an arbitrary scaling velocity and there is no upper limit for velocities. To chose between the non-positive definite pseudo-scalar product and the positive definite scalar product, we need some empirical facts, such as the finite speed of light c, which is the upper bound for velocities, or the relativistic aberration and Doppler shift. A derivation of the transformation based on aberration and Doppler shift shows that $V = \beta = v/c$ is given in Section 2.3.2, which confirms the correctness of the above assumptions.

Light cone and Minkowski distances

The square root of the pseudo-scalar product, Eq. (2.34), can be defined as the length of the 4-vector. If the pseudo-scalar product is zero, then

$$(c\,t)^2 = x^2 + y^2 + z^2\,. \tag{2.56}$$

This equation describes the propagation of a spherical wave with the speed of light in 3-dimensional Euklidian space. In 4-dimensional Minkowski space this is a cone, called the light cone. The light cone separates the Minkowski space into two parts, in which the pseudo-scalar product of a vector $\boldsymbol{r}$ of an event $\mathfrak{E}$ is either positive or negative. The positive part is called *time like* and the negative part *space like*, and the cone itself is *light like* (see Figure 2.10).

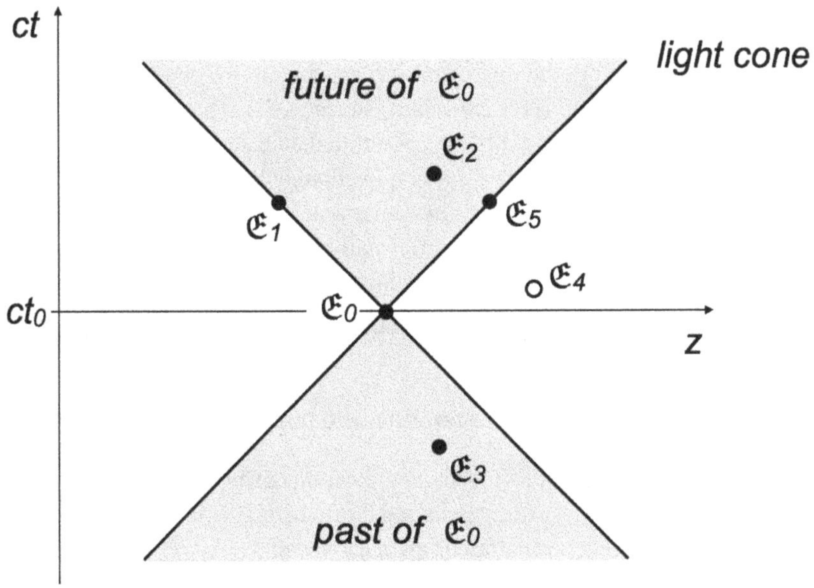

Fig. 2.10: Two dimensional 1+1 spacetime ct vs z plot of the light cone of event $\mathfrak{E}_0$ with coordinates $\boldsymbol{r}_0$. The five events $\mathfrak{E}_1$, $\mathfrak{E}_2$, $\mathfrak{E}_3$, $\mathfrak{E}_4$ and $\mathfrak{E}_5$ have a distinct relation to $\mathfrak{E}_0$. $\mathfrak{E}_1$ and $\mathfrak{E}_5$ lie on the light cone of $\mathfrak{E}_0$ and have no distance to $\mathfrak{E}_0$ in Minkowski space. $\mathfrak{E}_2$ lies in the future of $\mathfrak{E}_0$: $(\boldsymbol{r}_2 - \boldsymbol{r}_0)^2 > 0$ and $t_2 - t_0 > 0$ indicate that $\mathfrak{E}_0$ may cause an effect on $\mathfrak{E}_2$. $\mathfrak{E}_3$ lies in the past of $\mathfrak{E}_0$: $t_3 - t_0 < 0$ and $(\boldsymbol{r}_3 - \boldsymbol{r}_0)^2 > 0$ indicate that $\mathfrak{E}_0$ may feel an effect of $\mathfrak{E}_3$. $(\boldsymbol{r}_4\text{-}\boldsymbol{r}_0)^2$ is negative: the corresponding Minkowski distance is imaginary, which means that $\mathfrak{E}_4$ and $\mathfrak{E}_0$ are causally not connected, if they are not entangled.

The relation between *two* events may now be investigated. *Four* distinct relations are found: an event $\mathfrak{E}_A$ may be before, after, simultaneously, or causally not related to an other event $\mathfrak{E}_B$. To elaborate this, let $\mathfrak{E}_0$ and $\mathfrak{E}_i$ be distinct events with coordinates $\boldsymbol{r}_0$ and $\boldsymbol{r}_i$. The square of distance between $\mathfrak{E}_0$ and $\mathfrak{E}_i$ in Minkowski space is the pseudo-scalar product $(\boldsymbol{r}_i - \boldsymbol{r}_0) \cdot (\boldsymbol{r}_i - \boldsymbol{r}_0)$, and in 1+1 spacetime,

$$(\boldsymbol{r}_i - \boldsymbol{r}_0) \cdot (\boldsymbol{r}_i - \boldsymbol{r}_0) = c^2 \,(t_i - t_0)^2 - (z_i - z_0)^2 \,. \tag{2.57}$$

If this scalar product is zero, then the interval in spacetime or *Minkowski distance* between the two events $\mathfrak{E}_0$ and $\mathfrak{E}_i$ is zero. This means that the two events lie on the same light cone and are simultaneous, even if their temporal and spacial coordinates are distinct. If the pseudo-scalar product is positive, then its square root and distance is a real positive or negative number. If the pseudo-scalar product is negative, then its square root is an imaginary number. Such two events are causally not connected. Figure 2.10 illustrates distinct relations between events in Minkowski space.

Entanglement

The above statement that events with an imaginary distance in Minkowski space, such as it is that between $\mathfrak{E}_0$ and $\mathfrak{E}_4$ in Figure 2.10, are not causally connected can be challenged with entangled states. Entangled states consisting in two photons with the same polarization can be produced (Freedman and Clauser, 1972). For example, at $\mathfrak{E}_0$ a red and a green entangled x-polarized photon are emitted, one along z and the other along $-z$. If the red photon is detected at $\mathfrak{E}_1$, it is immediately known that there is a green x-polarized photon at $\mathfrak{E}_5$, even though the Minkowski distance between $\mathfrak{E}_1$ and $\mathfrak{E}_5$ is imaginary. This does not mean that the information on the green photon was transported at superluminal speed, which would violate causality, but the photon at $\mathfrak{E}_5$ was entangled with $\mathfrak{E}_1$ at $\mathfrak{E}_0$.

2.3.2 Lorentz transformations from aberration and Doppler shift

Aberration and Doppler shift of light give an indication of the transformation of space and time since they are related to directions and frequencies. In fact, Eqs. (2.15) and (2.23) are equivalent to the Lorentz transformation of a 4-vector $\boldsymbol{k} = (w/c, k_x, k_y, k_z)$, with w the angular frequency and (k_x, k_y, k_z) the direction of propagation. We note that the c is identified as the speed of light via the dispersion relation $c = \lambda\, \nu = w/k$, where $k = 2\pi/\lambda$ and $w = 2\pi\,\nu$. For a beam of light moving in the y-z plane and y'-z' plane, respectively, the 4-wave vector is

$$\boldsymbol{k} = \frac{w}{c}\begin{pmatrix} 1 \\ 0 \\ \sin\theta \\ \cos\theta \end{pmatrix}, \quad \boldsymbol{k}' = \frac{w'}{c}\begin{pmatrix} 1 \\ 0 \\ \sin\theta' \\ \cos\theta' \end{pmatrix}. \tag{2.58}$$

We assume that the transformation is described by a linear mapping

$$\boldsymbol{k}' = \boldsymbol{L}\cdot\boldsymbol{k}, \tag{2.59}$$

where $\boldsymbol{L}$ is the transformation matrix. With the assumption that only the direction parallel to the relative movement of the observer is affected, the matrix is written in the form

$$\boldsymbol{L} = \begin{pmatrix} a_{11} & 0 & 0 & a_{14} \\ 0 & 1 & 0 & 0 \\ 0 & 0 & 1 & 0 \\ a_{41} & 0 & 0 & a_{44} \end{pmatrix}. \tag{2.60}$$

The first two components of the vectors yield two equations for the matrix elements a_{11}, a_{14}, a_{41} and a_{44},

$$\begin{aligned} \nu\,(a_{11} + a_{14}\cos\theta) &= \nu' \,, \\ \nu\,(a_{41} + a_{44}\cos\theta) &= \nu' \cos\theta' \,, \end{aligned} \tag{2.61}$$

and finally with Eq. (2.23)

$$a_{11} = a_{44} = \gamma \,, \quad a_{14} = a_{41} = -\beta\gamma \,. \tag{2.62}$$

In many applications, where the relative motion of the observers is aligned with the first spacial coordinate (here z-axis), only the temporal part and the first spatial part are interesting,

$$\boldsymbol{L} = \begin{pmatrix} \gamma & 0 & 0 & -\beta\gamma \\ 0 & 1 & 0 & 0 \\ 0 & 0 & 1 & 0 \\ -\beta\gamma & 0 & 0 & \gamma \end{pmatrix} . \tag{2.63}$$

Although the 4-vectors of the light beams in Eqs. (2.58) have a ct-, y- and a z-component, the notation can be restricted to a 1+1 spacetime with corresponding 2×2 transformation matrices,

$$\boldsymbol{L} = \begin{pmatrix} \gamma & -\beta\gamma \\ -\beta\gamma & \gamma \end{pmatrix} , \tag{2.64}$$

which is the same result as in the previous Section 2.3.1. The inverse Lorentz transformation just changes the sign of the velocity β,

$$\boldsymbol{L}^{-1} = \begin{pmatrix} \gamma & \beta\gamma \\ \beta\gamma & \gamma \end{pmatrix} . \tag{2.65}$$

The results of this Section where aberration and Doppler shift is used to derive the Lorentz transformation of $\boldsymbol{k}$ vectors and thus the view of an Einsteinian observer extends the view of local observers as described in Section 2.2. In the mapping between the celestial spheres of two coinciding observers, measurements of distance do not play a role. On the other hand, the conformal mapping of the celestial sphere and the Doppler shift provide an absolute scale for distance measurements with the invariant speed of light. Radar measurements of distance, require a time measurement, for which sufficiently stable oscillators such as atoms or crystals are suitable. To be on the safe side with the definition of the measured time, the clock should move with the observer, thus measuring his proper time.

With such time and distance measurements, an observer can in principle define a co-moving spacetime coordinate system. The question concerning mapping one system to another arises if two observers want to communicate their observations. Yet, the line of thinking from the hypotheses about the conformal celestial spheres to the Lorentz transformation is not complete. The mathematical consistency justifies this approach, which finally must be confirmed by experience. In the given case, this link is provided by the introduction of the velocity, and by the radar method, which requires an operational time measurement, where both are taken from classical physics. The time measurement is restricted to proper time with a co-moving clock. The concluded Lorentz transformation finally shows the way, how the measurements can be interpreted in general.

2.3.3 Geometric representations of Lorentz transformations

Minkowski-Diagram

Hermann Minkowski (1909) introduced a diagram to illustrate the Lorentz transformation geometrically (Figure 2.11). This diagram handles the axes of the two coordinate systems in a asymmetric way, which requires different length and time scales in the two different systems. The z- and ct-axes are normal to each other, whereas the z'- and the ct'-axes are inclined. The angle α between the z-axis and the z'-axis is the same as that between the ct-axis and the ct'-axis. An event $\mathfrak{E}$ with coordinates $ct_{\mathfrak{E}}$ and $z_{\mathfrak{E}}$ is the Lorentz transform of $ct'_{\mathfrak{E}}$ and $z'_{\mathfrak{E}}$, and vice versa.

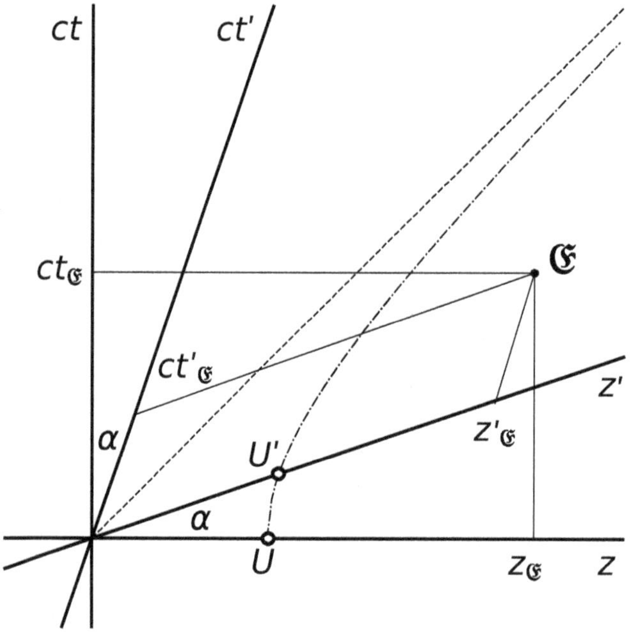

Fig. 2.11: Minkowski-diagram: $\mathfrak{E}$ is an event in spacetime. $(ct'_{\mathfrak{E}}, z'_{\mathfrak{E}})$ is the Lorentz transform of $(ct_{\mathfrak{E}}, z_{\mathfrak{E}})$ and vice versa. The dashed line is the world line of light emitted at the common origin (0,0). The hyperbola (dash-dotted line) is the position of all points with a distance of unity from the origin. U' is the unity of distance in the primed system.

The transformations between the primed and unprimed system in the Minkowski-diagram are

$$ct = ct' \cos\alpha + z' \sin\alpha \quad , \quad z = ct' \sin\alpha + z' \cos\alpha \,. \tag{2.66}$$

Compared to the Lorentz transformation, e.g. Eq. (2.65), this yields

$$\gamma = \cos\alpha \quad , \quad \beta\gamma = \sin\alpha \quad , \quad \beta = \tan\alpha \,, \tag{2.67}$$

thus the angle α is directly related to the velocity β.

All events with a distance of $U = 1$ from the origin of the systems are given by (see e.g. Eq. (2.34))

$$(ct)^2 - z^2 = 1 \,, \tag{2.68}$$

which is a hyperbola through the point (0,1) with the world line of light as asymptotic. The transformed Eq. (2.68) is

$$(ct')^2 - z'^2 = \frac{1}{\cos^2\alpha - \sin^2\alpha} = \frac{1 + \tan^2\alpha}{1 - \tan^2\alpha} \,. \tag{2.69}$$

Thus, the unit length U' ($z' = 1$) in the primed system of the Minkowski-diagram is

$$U' = U \sqrt{\frac{1 + \beta^2}{1 - \beta^2}} \,. \tag{2.70}$$

The unities U and U' are not the same in the two systems of the Minkowski-diagram.

Symmetric Minkowski-Diagrams: Loedel-Amar and Brehme

The disadvantages of the Minkowski-diagram, non-symmetric arrangement of the axes and different lengths of the unit distance in the diagram, can be overcome with the so called symmetric Minkowski-diagrams. These diagrams were proposed by Born (1920) and Gruner (1921). The axes of the symmetric diagrams are chosen such that the z-axis and the ct'-axis are orthogonal, the z'-axis and the ct-axis are orthogonal, and the angle between the z-axis and the z'-axis is α, see Figure (2.12). These diagrams are sometimes called Loedel-Amar-diagram (Loedel, 1948; Amar, 1955) (top of Figure 2.12) and Brehme-diagram (Brehme, 1961) (bottom of Figure 2.12), who reinvented them.

The transformation for these diagrams can be derived straightforwardly,

$$ct = ct' \cos\alpha \pm z \sin\alpha \quad , \quad z = z' \cos\alpha \pm ct \sin\alpha \,, \tag{2.71}$$

and with this we obtain

$$z = \frac{z' \pm ct' \sin\alpha}{\cos\alpha} \quad , \quad ct = \frac{ct' \pm z' \sin\alpha}{\cos\alpha} \,.$$

A comparison with Eq. (2.65) yields

$$\gamma = \frac{1}{\cos\alpha} \,, \ \beta\gamma = \pm\tan\alpha \ \text{ and } \ \beta = \pm\sin\alpha \,, \tag{2.72}$$

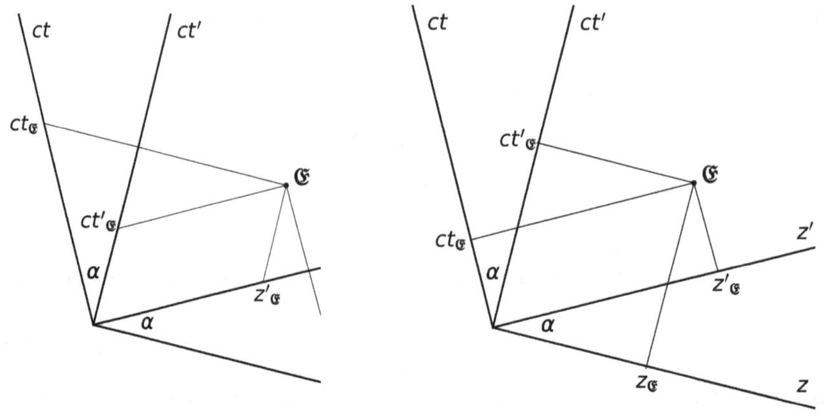

Fig. 2.12: Loedel-Amar-diagram (top panel) and Brehme-diagram (bottom panel). The coordinates $(ct'_{\mathfrak{E}}, z'_{\mathfrak{E}})$ and $(ct_{\mathfrak{E}}, z_{\mathfrak{E}})$ correspond to the Lorentz transformations of each other.

where the positive sign stands for the Loedel-Amar-diagram and the negative sign for the Brehme-diagram.

The transformed Eq. (2.68) in the symmetric diagrams is

$$z'^2 - c^2t'^2 = 1\,, \tag{2.73}$$

which is the same as in the unprimed systems. Here the unit length in the primed system is the same as in the unprimed system.

2.3.4 Algebraic representations of Lorentz transformations

The form of a Lorentz transformation corresponding to Eq. (2.65) is

$$\begin{pmatrix} ct \\ z \end{pmatrix} = \begin{pmatrix} f(\phi) & g(\phi) \\ g(\phi) & f(\phi) \end{pmatrix} \cdot \begin{pmatrix} ct' \\ z' \end{pmatrix}. \tag{2.74}$$

The general parameter ϕ for the movement is defined by the choice of the functions, and can be expressed with the coordinate velocity β,

$$\beta = g(\phi)/f(\phi)\,. \tag{2.75}$$

In most cases, no simple interpretation of ϕ is possible.

One example is inspired by the Brehme- and Loedel-Amar-diagrams in the previous section,

$$\begin{pmatrix} ct \\ z \end{pmatrix} = \begin{pmatrix} \dfrac{1}{\cos\alpha} & \dfrac{\sin\alpha}{\cos\alpha} \\[2ex] \dfrac{\sin\alpha}{\cos\alpha} & \dfrac{1}{\cos\alpha} \end{pmatrix} \cdot \begin{pmatrix} ct' \\ z' \end{pmatrix}, \tag{2.76}$$

with

$$\frac{1}{\cos\alpha} = \gamma\,, \quad \frac{\sin\alpha}{\cos\alpha} = \beta\gamma\,, \quad \sin\alpha = \beta, \tag{2.77}$$

where γ is the relativistic factor and β the coordinate velocity. The collinear velocity composition law is obtained by two consecutive transformations shown in Eq. (2.76)

$$\sin\alpha_3 = \frac{\sin\alpha_1 + \sin\alpha_2}{1 + \sin\alpha_1\,\sin\alpha_2}\,, \tag{2.78}$$

and is consistent with the velocity composition law in Eq. (2.31).

Another possibility is a function of a parameter σ, for which the composition law is multiplicative,

$$\begin{pmatrix} ct \\ z \end{pmatrix} = \begin{pmatrix} \dfrac{\sigma^2+1}{2\,\sigma} & \dfrac{\sigma^2-1}{2\,\sigma} \\ \dfrac{\sigma^2-1}{2\,\sigma} & \dfrac{\sigma^2+1}{2\,\sigma} \end{pmatrix} \cdot \begin{pmatrix} ct' \\ z' \end{pmatrix}. \tag{2.79}$$

where the relation of σ with the coordinate velocity is

$$\beta = \frac{\sigma^2-1}{\sigma^2+1}\,. \tag{2.80}$$

The composition of transformations yields that σ is multiplicative,

$$\sigma_{31} = \sigma_{12}\sigma_{23}\,. \tag{2.81}$$

And furthermore,

$$\sigma = \sqrt{\frac{1+\beta}{1-\beta}} \tag{2.82}$$

is the factor for the aberration, Eq. (2.15), and for the collinear Doppler shift, Eq. (2.24).

A third possible parameter is $\omega = \beta\,\gamma$, called the *proper velocity*, see Section 3.1.1, for which the transformation is

$$\begin{pmatrix} ct \\ z \end{pmatrix} = \begin{pmatrix} \sqrt{1+\omega^2} & \omega \\ \omega & \sqrt{1+\omega^2} \end{pmatrix} \cdot \begin{pmatrix} ct' \\ z' \end{pmatrix}. \tag{2.83}$$

Again, for ω no simple composition law follows,

$$\omega_3^2 = \omega_1\sqrt{1+\omega_2^2} + \omega_2\sqrt{1+\omega_1^2}\,. \tag{2.84}$$

For the high relativistic case, where $\omega \approx \gamma$, the composition of proper velocities simplifies to

$$\omega_3^2 = 2\,\omega_1\omega_2\,. \tag{2.85}$$

A fourth example uses the hyperbolic functions (Brehme, 1968),

$$\begin{pmatrix} ct \\ z \end{pmatrix} = \begin{pmatrix} \cosh\rho & \sinh\rho \\ \sinh\rho & \cosh\rho \end{pmatrix} \cdot \begin{pmatrix} ct' \\ z' \end{pmatrix}. \tag{2.86}$$

From Eqs. (2.63) and (2.86) follows

$$\beta = \tanh(\rho)\,. \tag{2.87}$$

The composition of transformations yields the additivity of the parameter,

$$\rho_3 = \rho_1 + \rho_2\,, \tag{2.88}$$

and thus the relation

$$\rho = \ln(\sigma) \tag{2.89}$$

to the multiplicative parameter σ. The additivity of ρ was mentioned by Pars (1921) and Taylor and Wheeler (1963). The similarity of the form (2.86) with the matrix for a rotation is sometimes used to *visualize* the Lorentz transformation as a hyperbolic rotation of Minkowski space. Taylor and Wheeler (1963) and Misner et al. (1973) called ρ the *velocity parameter*. Both, the parameter ρ as well as the parameter ω can be identified as operational measures of the velocity, as outlined in Section 3.1. The parameter ρ corresponds to the *rapidity* introduced in Eq. (2.9) and is further discussed in Section 3.1.2.

3 Consequences of the special theory of relativity

While "light" and the electrodynamics of moving bodies are embedded in special relativity, it was not obvious to translate special relativity to mechanics, where typical velocities are much smaller than the speed of light, and where classical Newtonian mechanics appears to describe the physics. This Chapter discusses the predictions and consequences of special relativity, starting with kinematics, discussing the corrections to Newtonian dynamics, and finally confirming the Lorentz invariance of Maxwellian electrodynamics.

3.1 Kinematics

In Minkowski's world, space and time, and thus motion, depend on the state of motion of an observer, in contrast to a classical world, where e.g. all observers measure the same time interval between two given events. Therefore in special relativity, a generalization of physical quantities such as velocity and acceleration allows for more than one possible definition of "velocity", and each of them may be useful. This requires a carefully defined terminology.

3.1.1 Coordinate velocity and proper velocity

Motion is generally described with a velocity, i.e. the differential limit

$$v = \lim_{\Delta t \to 0} \frac{\Delta s}{\Delta t} , \tag{3.1}$$

where Δs is a change in position in space within the time step Δt. In Galilean spacetime, Δt and Δs are the same in all frames of reference. In relativistic spacetime this is not the case and a careful distinction between different operational definitions of velocity must be made. It is important to distinguish whether the two quantities Δs and Δt are measured in the same rest frame or not. Here, we refer to the scenario in Figure 2.5.

If both, distance and time are measured in the same system of reference, O or O', we get the coordinate velocity $\beta = v/c$,

$$\beta = \frac{1}{c} \lim_{\Delta t \to 0} \frac{\Delta s}{\Delta t} \quad , \qquad \beta' = \frac{1}{c} \lim_{\Delta t' \to 0} \frac{\Delta s'}{\Delta t'} , \tag{3.2}$$

where the velocities in the two systems are opposite $\beta = -\beta'$.

If distance and time are measured in different frames of reference, we obtain the *proper velocity*, sometimes called *celerity*. This is e.g. applied by car drivers for the calibration of their tachometers, where the rest frame 'car' is frame O' and that of the street frame O. The time step $\Delta t'$ is measured on the clock in the car, and the change in position

https://doi.org/10.1515/9783111503592-004

(kilometer posts) Δs is measured in the rest frame of the street. The resulting quantity is the proper velocity,

$$\omega = \frac{1}{c} \lim_{\Delta t' \to 0} \frac{\Delta s}{\Delta t'} \,. \tag{3.3}$$

The reciprocal measurement yields the same result: an observer outside the car measures the time Δt on her clock between the passages of the front and rear ends of the car, using the proper length $\Delta l = \Delta s'$ of the car as specified by the manufacturer,

$$\omega' = \frac{1}{c} \lim_{\Delta t \to 0} \frac{\Delta s'}{\Delta t} \,. \tag{3.4}$$

There is some ambiguity concerning the definition of the sign of the proper velocity. If the length of the car is defined as a positive number, then ω' is also a positive number and refers to the velocity of the car relative to the street. If we stick to the scenario in Figure 2.5, which is preferable, then $\Delta s'$ is negative, thus ω' is negative and refers to the velocity of the street relative to the car. Since $\Delta t = \gamma \, \Delta t'$ and $\Delta s' = \gamma \Delta s$, the proper velocities are related to the coordinate velocity by

$$\omega = -\omega' = \gamma \beta \,. \tag{3.5}$$

In contrast to β the proper velocity ω is an unbound quantity. Substituting Eq. (3.5) in Eq. (2.18) yields for the relativistic factor

$$\gamma = \sqrt{1 + \omega^2} \,. \tag{3.6}$$

Approaching the speed of light $\beta \to 1$, i.e. for $\omega \gg 1$

$$\omega \approx \gamma \,. \tag{3.7}$$

In Figure 3.1 the above relations of the different velocities is summarized in a table.

	Δt	$\Delta t'$
Δs	β	ω
$\Delta s'$	$-\omega$	$-\beta$

Fig. 3.1: Velocities depend on the change in position Δs or $\Delta s'$ and the time intervals Δt or $\Delta t'$, which can be measured in two systems that move relative to each other. If the two differential quantities are not measured in the same reference frame, the *coordinate velocity* β has to be multiplied with the relativistic factor γ, which yields the *proper velocity* ω.

3.1.2 Rapidity, the integrated proper acceleration

A third possibility to quantify motion is using quantities measured in a frame without connection to other frames of reference such as it is encountered in a spacecraft without windows. The time measurement within this frame is commonly called *proper time* τ. As the definition of a velocity relies on a comparison to a reference system (Section 3.1.1), the velocity of a single frame is not defined. A change in velocity can be recorded as an acceleration which involves different frames of reference. Accordingly we define the change in *rapidity* $\Delta\rho$ as

$$\Delta\rho = \alpha\,\Delta\tau\,, \tag{3.8}$$

where we call α *proper acceleration*. As β, ω and ρ are dimensionless, α has the dimension of a rate [s^{-1}]. The integral of the *proper acceleration* with proper time gets (Levy-Leblond and Provost, 1979; Levy-Leblond, 1980),

$$\rho(\tau) = \int_0^{\tau} \alpha(\tau')\,\mathrm{d}\tau' + \rho_0\,. \tag{3.9}$$

This determines the rapidity up to the a priori unknown integration constant or initial rapidity ρ_0.

The measurement of acceleration is not critical in classical physics, where it is defined by

$$\alpha = \frac{1}{c}\lim_{\Delta t\to 0}\frac{\Delta v}{\Delta t}\,. \tag{3.10}$$

From Section 3.1.1 we know that in relativistic physics this needs further analysis. For relativistic velocities, several possibilities exist. The change in coordinate velocity $\Delta\beta$, proper velocity $\Delta\omega$ or rapidity $\Delta\rho$ can be considered. Also care has to be taken in which frame the time is measured. In Figure 3.2 the concept of the three different relativistic speeds is outlined in a matrix.

In the following we want to express the proper acceleration

$$\alpha = \lim_{\Delta\tau\to 0}\frac{\Delta\rho}{\Delta\tau} \tag{3.11}$$

with the coordinate velocity β and the proper velocity ω. To express the proper acceleration with the coordinate velocity of an arbitrary inertial frame we define 3 different frames of reference and apply the relativistic velocity composition law (Section 2.2.5). Frame 1 and 2 correspond to the inertial frames before and after proper acceleration. The change in rapidity is therefore $\Delta\rho = \beta_{12}$, where frame 2 moves with β_{12} in frame 1 and frame 1 with $\beta_{21} = -\beta_{12}$ in frame 2. Frame 3 is the arbitrary frame that moves with β_{13} with respect to frame 1 and with β_{23} with respect to frame 2. $\beta_{23} - \beta_{13} = -\Delta\beta$ is the change in coordinate velocity of frame 3 in going from frame 1 to frame 2, i.e. during the proper acceleration. With the velocity composition law Eq. (2.31), we find

$$\Delta\rho = \beta_{12} = \frac{\beta_{13} - \beta_{23}}{1 - \beta_{13}\beta_{23}} = \frac{\Delta\beta}{1 - \beta_{13}(\beta_{13} - \Delta\beta)} \tag{3.12}$$

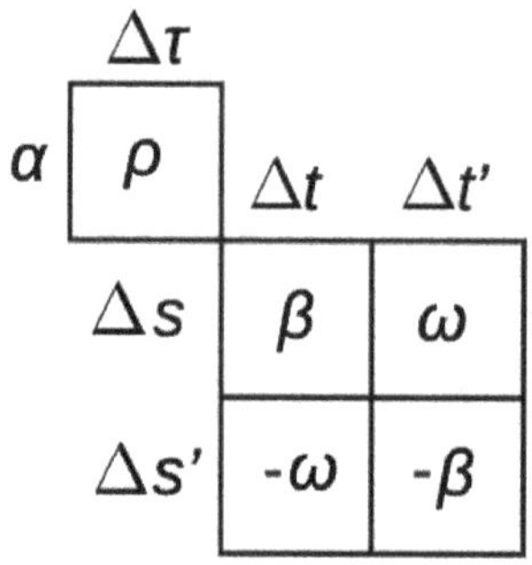

Fig. 3.2: Concept of the three relativistic velocities β, ω and ρ. If the acceleration is integrated in the *proper time* τ of an accelerating system the rapidity ρ is obtained. Like the *proper velocity* ω, the rapidity ρ is an unbound quantity.

In the differential limit for vanishing length of the time interval $\Delta\tau$ and thus $\Delta\beta \to 0$,

$$\mathrm{d}\rho = \gamma^2\,\mathrm{d}\beta \tag{3.13}$$

follows. The proper acceleration α defined by the velocity increment $\mathrm{d}\rho$ and the proper time interval $\mathrm{d}\tau$ expressed with $\mathrm{d}\beta$ gets

$$\alpha = \frac{\mathrm{d}\rho}{\mathrm{d}\tau} = \gamma^2\,\frac{\mathrm{d}\beta}{\mathrm{d}\tau}\,. \tag{3.14}$$

With this we solve Eq. (3.9) for a scenario where at $\tau = 0$ a point or a spacecraft is accelerated with α,

$$\rho = \int_0^{\tau} \alpha\,\mathrm{d}\tau' + \rho_0 = \int_0^{\beta} \gamma^2\,\mathrm{d}\beta' + \rho_0 = \operatorname{arctanh}(\beta) + \rho_0\,. \tag{3.15}$$

From Eq. (3.15) it can be seen that rapidity is an unbound quantity as it is the proper velocity. Comparing Eq. (3.15) with Eqs. (2.75) and (2.86), we find with the choice for the initial frame of reference $\rho_0 = 0$

$$\gamma = \cosh(\rho) \quad \text{and} \quad \omega = \gamma\,\beta = \sinh(\rho)\,. \tag{3.16}$$

With Eq. (3.16), the proper acceleration α can be expressed as

$$\alpha = \frac{\mathrm{d}\rho}{\mathrm{d}\tau} = \frac{\mathrm{d}}{\mathrm{d}\tau}\operatorname{arcsinh}(\omega) = \frac{1}{\sqrt{1+\omega^2}}\,\frac{\mathrm{d}\omega}{\mathrm{d}\tau} = \frac{\mathrm{d}\omega}{\mathrm{d}t} = \gamma^3\,\frac{\mathrm{d}\beta}{\mathrm{d}t}\,, \tag{3.17}$$

where $\mathrm{d}\beta/\mathrm{d}t$ is the distinct coordinate acceleration. This is an interesting and useful result. The proper acceleration α is not only the derivative of rapidity ρ with respect to the proper time τ, but it is equal to the derivative of the proper velocity ω with respect to coordinate time t. The definition of proper acceleration corresponds to the spatial part of the relativistic acceleration 4-vector.

Eqs. (3.17) suggest to define a constant acceleration as constant proper acceleration, $\mathrm{d}\omega/\mathrm{d}t$, rather than a constant coordinate acceleration, $\mathrm{d}\beta/\mathrm{d}t$. For constant proper acceleration, the increase of proper velocity with proper time is given by

$$\frac{\mathrm{d}\omega}{\mathrm{d}t} = \frac{\mathrm{d}\omega}{\gamma\,\mathrm{d}\tau} = \alpha\,. \tag{3.18}$$

This equation can be separated by variables,

$$\frac{\mathrm{d}\omega}{\sqrt{1+\omega^2}} = \alpha\,\mathrm{d}\tau\,, \tag{3.19}$$

and with $\omega = 0$ at $\tau = 0$ the solution of the integral is

$$\ln\left(\omega + \sqrt{1+\omega^2}\,\right) = \alpha\,\tau\,. \tag{3.20}$$

For high relativistic speeds, $\omega \gg 1$ and a constant α, the proper velocity grows exponentially with proper time,

$$\omega = \frac{1}{2}\,\mathrm{e}^{\alpha\,\tau}\,. \tag{3.21}$$

The path length of an accelerated point in an inertial reference frame, such as it may be encountered in space travel is straight forward to calculate. Consider a motion with constant proper acceleration α with the velocity $\beta_0 = \omega_0 = \rho_0 = 0$ at time $t_0 = \tau_0 = 0$. Then the rapidity at proper time τ is $\rho = \alpha\tau$, and with Eq. (3.3), the distance traveled in an inertial frame of reference is

$$s = \int_0^s \mathrm{d}s' = c\int_0^\tau \omega(\tau')\,\mathrm{d}\tau' = c\int_0^\tau \sinh(\alpha\tau')\,\mathrm{d}\tau' = \frac{c}{\alpha}\,[\cosh(\alpha\tau) - 1]\,. \tag{3.22}$$

In Section 7.1.2 we will further explore this equation and its consequences on clocks in the frame of reference at the beginning of a space trip and that at τ on an uniformly accelerated spacecraft.

3.2 Dynamics

The previous Section dealt with kinematics, i.e. with points in space and time, motion and change of motion, without asking for a reason for motion nor its change. In this Section, we consider the implications of special relativity to dynamics. To do this, we adopt Newton's concept of forces acting on objects with inertial mass.

3.2.1 Newton's second law

Newton's laws (see Section 1.4.1) also hold in the relativistic context although some of the elements have to be adopted. The first law or principle of inertia states that massive bodies remain at rest or in uniform linear motion as long as there is no force acting on them. This has no implications for the transition from Newtonian to Einsteinian dynamics. Also the third law, actio=reactio, is not affected by special relativity. The second principle, mostly referred to as *Newton's second law*, $F=m\,a$ in Eq. (1.1) requires a revision because it involves space and time and more than one frame of reference. This has consequences on the dynamics. We have to carefully consider the kinematic implications on the acceleration. The case of linear and circular motion will be treated. In linear motion, the force acts parallel to the velocity, and in circular motion it acts perpendicular to the velocity.

Linear motion

Newton's second law is $F = \mathrm{d}(m\,v)/\mathrm{d}t$, where F is the force acting on the mass m with momentum mv during the time increment dt, where v is the velocity. In Einsteinian dynamics, it is rewritten as

$$F = \frac{\mathrm{d}}{\mathrm{d}t}\left(\gamma\, m\, v\right) \tag{3.23}$$

and it has to be specified that the time increment $\mathrm{d}t$ is measured in the frame of reference in which the mass is moving with velocity v. In Eq. (3.23) all physical quantities are defined in the same frame of reference.

In this Section, a derivation of relativistic dynamics is suggested in which Newton's second law is written in quantities described in the rest frame of the accelerated mass. These quantities are the invariant rest mass, the proper time, and the rapidity (see Section 3.1.2). First, the derivation is restricted to linear motion and collinear acceleration.

Rapidity needs a concept of inertia for the measurement of acceleration. An accelerometer measures acceleration independent of the actual speed with respect to any inertial frame of reference. If this were not the case, then one specific inertial frame could be singled out to serve as an absolute frame of reference, and the principle of relativity would be violated. A measurement of weight in a laboratory or in a spacecraft flying in space is a static measurement with respect to the rest frame of the laboratory or the spacecraft, and Newton's second law applies exactly. An astronaut feeling a constant weight thus correctly interprets the motion as uniformly accelerated. With a gyroscope, it is possible to distinguish between a linearly accelerated or purely circular motion or any combination of both.

In the simplest case of linear acceleration, the astronaut interprets the acceleration as a change of the rapidity per proper time interval. Furthermore, the constant weight of the astronaut is correctly interpreted as the constant force needed to accelerate the rest mass m, independent of the rapidity. In the case of non-uniform linear acceleration, the momentary weight F can be interpreted as the constant force needed to accelerate the mass m of the astronaut with the acceleration $c\,\alpha$ (see Eqs. (3.10) and (3.17)), again independent of the rapidity,

$$F = m\,c\,\alpha = m\,c\,\frac{\mathrm{d}\rho}{\mathrm{d}\tau} = m\,c\,\frac{\mathrm{d}\omega}{\mathrm{d}t} = c\,\frac{\mathrm{d}}{\mathrm{d}t}(\gamma\, m\,\beta) = m\,c\,\gamma^3\,\frac{\mathrm{d}\beta}{\mathrm{d}t}, \tag{3.24}$$

which is the relativistic "Second Law", Eq. (3.23), written in terms of velocity and time as measured in an inertial frame of reference moving with velocity $v = c\,\beta$ relative to an astronaut with mass m. Note that the factor γ^3 is purely kinematic, see also Eq. (3.17). This makes it clear that the factor γ^3 in the relativistic relation between force and acceleration stems from the transformation of space and time. A constant Force:mass ratio does not alter the findings in dynamics from those in kinematics.

Circular motion

A similar path of considerations can be applied to determine the relationship between force and radius of a circular path or trajectory if a force acts transverse to the direction of motion. We use the idea that a movement can be decomposed into a number of independent components by means of a linear combination of vectors. Circular motion is locally defined by a tangential uniform motion and a radially accelerated motion. Thus, the local relation between the tangential velocity, the radial acceleration, and the radius of the locally circular path can be obtained as an infinitesimal limit of the superposition of the two components of motion.

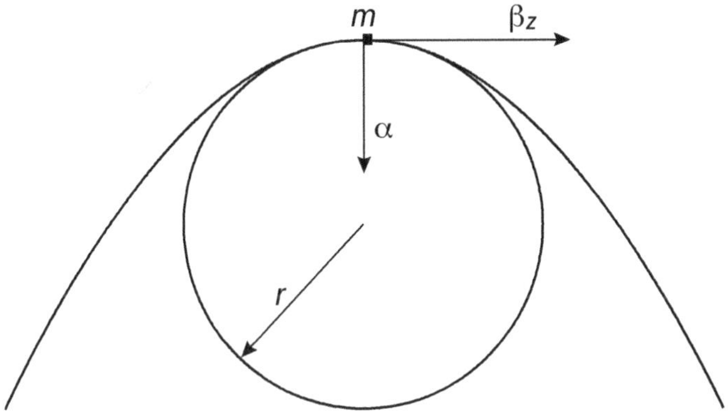

Fig. 3.3: Transverse acceleration of a mass m: a linear uniform motion in the z-direction with velocity β_z and a transverse proper acceleration a_r yield locally a parabolic trajectory with an osculating circle with radius r. At the vertex of the parabola (y, z)=(0, 0) the transversal velocity is β_y=0.

We consider a particle (spacecraft) to move uniformly in one direction while it is accelerated in another direction. Let the mass m move with proper velocity ω at time $t = \tau = 0$ through the origin $(y, z) = (0, 0)$ in the direction of the positive z-axis, thus

$$z = \omega\, c\, \tau\,. \tag{3.25}$$

A force F_r in the direction of the positive y-axis accelerates the mass in this direction with proper acceleration,

$$a_r = \frac{F_r}{m\,c} = \gamma^3\,\frac{\mathrm{d}\beta}{\mathrm{d}t}\,. \tag{3.26}$$

see Eq. (3.24). Its position y at proper time τ is given by Eq. (3.22),

$$y = \frac{c}{a_r}\,[\cosh(a_r\tau) - 1]\,. \tag{3.27}$$

Eqs. (3.25) and (3.27) constitute a parameter equation for the path of a relativistic object that is accelerated perpendicular to its motion. For a circular motion, the acceleration must be normal to the direction of the momentary motion at every time. At the moment when the spacecraft passes the coordinate origin $(y, z) = (0, 0)$, the conditions for a circular motion are given by the acceleration a, the proper velocity ω and the radius of the osculating circle of the above trajectory, Eqs. (3.25) and (3.27).

The Taylor expansion of $\cosh(a\tau)$, truncated after the second order term, yields the local trajectory of the radially accelerated movement around the starting point $(0, 0)$,

$$y = \frac{a_r\,c}{2}\,\tau^2\,, \tag{3.28}$$

which is in fact the non-relativistic approximation of a uniformly accelerated motion, and is always true in an infinitesimal surrounding of the starting point. Eqs. (3.25) and (3.28) define a very good parametric representation of the path near the origin $(0, 0)$, which by elimination of τ yields an explicit equation for a parabola with a vertex at the origin (Figure 3.3),

$$y = \frac{a_r}{2\,c\,\omega^2}\,z^2\,. \tag{3.29}$$

The radius of the osculating circle of the parabola is

$$r = c\,\frac{\omega^2}{a_r}\,, \tag{3.30}$$

from which we obtain the centripetal acceleration a_r for a circular motion with radius r,

$$a_r = c\,\frac{\omega^2}{r} \tag{3.31}$$

and with Eq. (3.24) the relativistic equation for the centripetal force for a circular motion of a mass m,

$$F_r = \frac{m\,c^2\,\omega^2}{r} = \frac{m\,c^2\,\gamma^2\,\beta^2}{r} = \frac{m\,\gamma^2\,v^2}{r}\,. \tag{3.32}$$

As for the above derived relativistic *second law* for linear motion, the factor γ stems from the transformation of space and time due to the tangential motion and is not influenced by radial motion, which is always infinitesimally small on the entire circular orbit.

For circular motion, it is useful to introduce an angular velocity analogous to classical kinematics. For uniform circular motion, the proper velocity is

$$\omega = \frac{1}{c}\,\frac{\Delta l}{\Delta\tau} = \frac{2\pi\,r}{c\,T_\mathrm{p}}\,, \tag{3.33}$$

where T_p is the proper time period for one revolution. The proper angular velocity thus becomes

$$\Omega_p = \frac{2\pi}{T_p} = \gamma \frac{2\pi}{T_k} = \gamma\, \Omega_k\,, \tag{3.34}$$

where $T_k = \gamma\, T_p$ are the times for one revolution in the accelerated (T_p) and the non-accelerated (T_k) system. Ω_k is the angular velocity in the non-accelerated coordinate system of the center of the circle. The above relativistic centripetal force written in proper angular velocity corresponds to the classical centripetal force written in coordinate angular velocity,

$$F_r = m\, r\, \Omega_p^2\,. \tag{3.35}$$

A posteriori we note that relativistic circular motion is described like non-relativistic circular motion within kinematics (Section 3.1). A constant force:mass ratio does not alter the findings.

3.2.2 Momentum and energy

Using the proper quantities, the notation of classical kinematics can be recovered for relativistic kinematics, Eqs. (3.4) and (3.17), such that the proper velocity is the derivative of proper length with respect to coordinate time and acceleration is the derivative of proper velocity with respect to coordinate time. This pattern can be extended to dynamics by defining the force F as the derivative of linear momentum p with respect to coordinate time. From Eq. (3.24) we get

$$F = m\, c\, \frac{\mathrm{d}\omega}{\mathrm{d}t} \equiv \frac{\mathrm{d}p}{\mathrm{d}t}\,, \tag{3.36}$$

and consequently, the linear momentum gets with Eqs. (3.5) and (3.6)

$$p = m\, c\, \omega = m\, c\, \gamma\, \beta = m\, c\, \sqrt{\gamma^2 - 1}\,, \tag{3.37}$$

expressed in the proper velocity ω, the coordinate velocity β or the relativistic factor γ.

The kinetic energy of a relativistic particle is best derived by computing the work applied to a particle to accelerate it to a velocity β (Tipler, 1999),

$$E_{\mathrm{kin}} = \int_0^d F\, \mathrm{d}s\,, \tag{3.38}$$

where the particle travels a distance d in the system where E_{kin} is measured. Applying the relativistic second law in Eq. (3.24), we obtain

$$E_{\mathrm{kin}} = m\, c \int_0^d \gamma'^3 \frac{\mathrm{d}\beta'}{\mathrm{d}t}\, \mathrm{d}s = m\, c^2 \int_0^\beta \beta'\, \gamma'^3\, \mathrm{d}\beta' = m\, c^2\, (\gamma - 1)\,, \tag{3.39}$$

where the dashed variables β' and γ' denote the integration variables, and the kinetic energy E_{kin} is a function of the final γ. This is perhaps Einsteins most famous equation (Einstein, 1905a). Eq. (3.39) contains a term that depends with γ on the motion and on a constant term, which is the rest energy of the mass m in its rest frame

$$E_0 = m\,c^2\,, \tag{3.40}$$

This equation implies that mass and energy are equivalent and that mass can be transformed into energy and vice versa.

From Eq. (3.39) and the first two non-vanishing terms of the Taylor expansion of γ, which are $1 + \beta^2$, we obtain the familiar result for the non-relativistic kinetic energy $(\beta \to 0)$

$$E_{\mathrm{kin}} \approx \frac{1}{2} m v^2\,. \tag{3.41}$$

The total energy E_{tot} that is conserved in an isolated system and invariant under Lorentz transformations is:

$$E_{\mathrm{tot}} = E_0 + E_{\mathrm{kin}} = \gamma\, m c^2\,. \tag{3.42}$$

With Eqs. (3.37) and (3.39), the momentum expressed in terms of the kinetic energy is

$$p = \sqrt{E_{\mathrm{kin}}^2 + 2E_{\mathrm{kin}}\, m\, c^2}\,, \tag{3.43}$$

and correspondingly for the coordinate velocity β and proper velocity ω,

$$\beta = \frac{\sqrt{E_{\mathrm{kin}}^2 + 2E_{\mathrm{kin}}\, m\, c^2}}{E_{\mathrm{kin}} + m\, c^2}\,, \qquad \omega = \frac{\sqrt{E_{\mathrm{kin}}^2 + 2E_{\mathrm{kin}}\, m\, c^2}}{m\, c^2}\,. \tag{3.44}$$

3.3 Electrodynamics

Electrodynamics is a genuine relativistic theory that has no classical (Galilean) counterpart such as mechanics with its classical and relativistic versions. The work of Maxwell (1873) is the first comprehensive treatment of electrodynamics. The equations, now named Maxwell's equations can be found in this work. They were condensed later in four vector field equations by Heavyside in 1884 (Nahin, 1988) and they were named Hertz-Heaviside equations, Maxwell-Hertz equations or Maxwell-Heaviside equations, before they got today's label "Maxwell equations".

3.3.1 Maxwell equations

The Maxwell equations are Lorentz invariant and thus, implicitly contain the special theory of relativity. To elaborate this we recall the Maxwell equations in the International System of Units (SI units) in vector form:

the Gauss law for electric fields,

$$\nabla \cdot \mathbf{E} = \frac{\rho_q}{\epsilon_0}\,, \tag{3.45}$$

the Gauss law for magnetic fields,

$$\nabla \cdot \mathbf{B} = 0\,, \tag{3.46}$$

the Maxwell-Faraday equation,

$$\nabla \times \mathbf{E} = -\frac{\partial \mathbf{B}}{\partial t}\,, \tag{3.47}$$

and the law of Ampère,

$$\nabla \times \mathbf{B} = \frac{1}{\epsilon_0\, c^2}\,\mathbf{j} + \frac{1}{c^2}\,\frac{\partial \mathbf{E}}{\partial t}\,, \tag{3.48}$$

where $\mathbf{E}(x, y, z, t)$ is the electric field, $\mathbf{B}(x, y, z, t)$ the magnetic field, ρ_q the volume density of the electric charge, $\mathbf{j}$ the electric current density, ϵ_0 is the vacuum permittivity and c is the speed of light. ∇ is the Nabla differential operator $(\frac{\partial}{\partial x}, \frac{\partial}{\partial y}, \frac{\partial}{\partial z})$, thus $\nabla\cdot$ denotes the divergence and $\nabla\times$ the curl of a given vector field. The Maxwell equations for $\mathbf{E}(x, y, z, t)$ and $\mathbf{B}(x, y, z, t)$ are linear coupled differential equations, where solutions may be added or superimposed in linear combinations. The differential operators in the Maxwell equations do not exclude constant fields in space and time to be solutions. If such constant fields were not zero, this would imply infinite energy, which we neglect in the following.

Implications and consequences

- If the speed of light were infinite we would expect no magnetic fields. This indicates the relativistic character of electrodynamics and the fact that magnetic fields are inseparably connected to the finite and invariant speed of light.
- The current density $\mathbf{j}$ corresponds to the flux of charge density,

  $$\mathbf{j} \equiv \rho_q\, \mathbf{v}\,, \tag{3.49}$$

  where $\mathbf{v}$ is the drift velocity of the charge density. In a static scenario $\partial \mathbf{E}/\partial t = 0$, and in a frame of reference without moving charges ($\mathbf{j} = 0$) no $\mathbf{B}$-fields emerge.
- Charge conservation follows from the divergence of Eq. (3.48). With Eq. (3.45) and with the identity $\nabla \cdot (\nabla \times \mathbf{B}) = 0$ we get

  $$\nabla \cdot \mathbf{j} + \frac{\partial \rho_q}{\partial t} = 0\,. \tag{3.50}$$

Solutions in the vacuum: light

The Maxwell equations in the vacuum, i.e. without charge ($\rho_q(x, y, z) = 0$) simplify to

$$\nabla \cdot \mathbf{E} = 0\,, \tag{3.51}$$

$$\nabla \cdot \mathbf{B} = 0\,, \tag{3.52}$$

$$\nabla \times \mathbf{E} = -\frac{\partial \mathbf{B}}{\partial t}\,, \tag{3.53}$$

$$\nabla \times \mathbf{B} = \frac{1}{c^2}\frac{\partial \mathbf{E}}{\partial t}\,. \tag{3.54}$$

Electromagnetic waves including visible light with a wavelength and frequency, propagating with the speed of light are the solutions. Except for signs and constant factors, this set of coupled homogeneous differential equations is symmetric in the **E** and **B** fields. Eqs. (3.51) to (3.54) have an infinite set of solutions besides $\mathbf{E} = 0$ and $\mathbf{B} = 0$ as the trivial solutions.

To show this we apply $\nabla\times$ to Eq. (3.53), differentiate Eq. (3.54) with respect to time t, add the equations and apply the identity $\nabla \times (\nabla \times \mathbf{E}) = \nabla(\nabla \cdot \mathbf{E}) - \Delta\mathbf{E}$ to obtain a wave equation for the Laplacian Δ of the electric field,

$$\Delta\mathbf{E} = \frac{1}{c^2}\frac{\partial^2 \mathbf{E}}{\partial t^2}\,. \tag{3.55}$$

Equivalently, we apply $\nabla\times$ to Eq. (3.54), differentiate Eq. (3.53) with respect to time t, add the equations and apply the identity $\nabla\times(\nabla\times\mathbf{B}) = \nabla(\nabla \cdot \mathbf{B}) - \Delta\mathbf{B}$ to obtain a wave equation for the magnetic field,

$$\Delta\mathbf{B} = \frac{1}{c^2}\frac{\partial^2 \mathbf{B}}{\partial t^2} \tag{3.56}$$

These are wave equations with plane waves as solutions,

$$\mathbf{E} = \mathbf{E}_0\, e^{i(wt - \mathbf{k}_s \cdot \mathbf{r}_s)}\,, \tag{3.57}$$

$$\mathbf{B} = \mathbf{B}_0\, e^{i(wt - \mathbf{k}_s \cdot \mathbf{r}_s)}\,, \tag{3.58}$$

where $\mathbf{E}_0=(E_x, E_y, E_z)$ and $\mathbf{B}_0=(B_x, B_y, B_z)$ are the amplitudes of the electric and the magnetic fields, respectively, and $i=\sqrt{-1}$. $\mathbf{k}_s$ indicates the *wave vector* along the propagation direction $\mathbf{r}_s$. $w/2\pi$ is the frequency ν and $2\pi/k$ the wavelength λ.

Inserting the solutions into the equations, i.e. Eq. (3.57) into Eq. (3.55) and Eq. (3.57) into Eq. (3.56), confirms the solutions, and yields the *dispersion relation* for electromagnetic radiation

$$c = \lambda\nu = \frac{w}{k}\,. \tag{3.59}$$

Since Eqs. (3.55) and (3.56) are linear differential equations, any linear combination of particular solutions is a solution, too.

To understand the properties of the solutions, it is sufficient to discuss the properties of one particular solution. From this we will see on how **E**, **B** and $\mathbf{k}_s$ are related for a particular solution: Eqs. (3.53) and (3.54) impose further constraints for $\mathbf{E}_0$ and $\mathbf{B}_0$. From Eq. (3.53) we conclude that a change of the magnetic field only concerns a change in the component normal to the electric field, and vice versa for the electric field from Eq. (3.54). This implies that the variable parts of the components of the **E** and **B** fields must be normal to each other at any given point in space and time. The orientation of the wave vector $\mathbf{k_s}$ is inferred from the first two Maxwell equations Eqs. (3.51) and (3.52), from where it follows that $\mathbf{k_s}$ is normal to **E** and **B** at every point in space and time. This confirms the picture of electromagnetic radiation as a transversal waves with an isotropic propagation velocity that corresponds to the speed of light c.

Now we want to discuss the *polarisation* of electromagnetic radiation. Above we have seen for a given solution that **E** and **B** are orthogonal and both are orthogonal to the wave vector $\mathbf{k_s}$. Though, we did not state in which direction **E** and **B** point within the plane perpendicular to $\mathbf{k_s}$. The solutions (3.57) and (3.58) have to be specified with two more parameters φ and $\dot{\varphi}$. Without loss of generality, we investigate an electromagnetic wave propagating along the z axis and we treat the electric field amplitude only. For this scenario the **E** vector lies in the x-y plane, has components $(E_x, E_y, 0)$ and the magnitude $E_0=\sqrt{E_x^2+E_y^2}$. The orientation of the **E** vector is best expressed with polar coordinates $E_x=E_0\cos(\varphi)$ and $E_y=E_0\sin(\varphi)$, where $\varphi=0$ stands for x-polarisation and $\varphi=\pi/2$ for y-polarisation. For constant φ the polarisation is linear. If φ changes with time we obtain circular or elliptic polarisation, where the **E** vector rotates in the x-y plane with $\varphi(t)=\varphi_0+\dot{\varphi}\,t$. The rotation rate $\dot{\varphi}$ is $\pm w$ for left or right circularly polarized light. Notably $\dot{\varphi}$ is 0 or $\pm w$ if the periodicity condition $\mathbf{E}(x,y,z,t)=\mathbf{E}(x,y,z,t+T)$, with the period $T=2\pi/w$ is fulfilled.

Lorentz invariance of the Maxwell equations

The Maxwell equations were first shown to be Lorentz invariant by Poincaré (1905). Here we do not prove the invariance for the Maxwell equations themselves, but we show the invariance of their solutions in vacuum Eqs. (3.57) and (3.58). In different frames of reference the dispersion relation Eq. (3.59) remains, up to a change in frequency, and wave vector, the same (see Sections 1.2.1 and 2.3.2): Electromagnetic radiation remains electromagnetic radiation. Importantly, the electromagnetic radiation propagates in all reference frames with the same speed c and we find for the dispersion relation in Eq. (3.59),

$$c=\frac{w}{k}=\frac{w'}{k'}\,. \tag{3.60}$$

Since both, the angular frequency w and the wavenumber k are equally Doppler shifted, Lorentz invariance is demonstrated. We note that the aberration, i.e. the change in the apparent propagation *direction* is implicitly considered in the Doppler shift.

3.3.2 Lorentz transformation of electromagnetic fields

The transformation of electric and magnetic fields from one to another frame of reference reveals properties of the fields and the relativistic nature of electrodynamics. For this purpose, one of the simplest non-trivial situations is treated in detail.

Consider a homogeneous linear line of charge along the z-axis. If this line of charge is at rest it displays an electrostatic field only. If it is in motion a magnetic field emerges. This scenario is simpler than a moving point charge as it defines a cylindrical symmetry along the z-axis and reduces the problem to two dimensions. We discuss two frames of reference O and O', where O is the laboratory frame in which the charge is moving with β along the z axis, and O' the rest frame of the charge line. The charge line densities are ρ_{q1} and ρ'_{q1}, respectively. Due to the Lorentz contraction (see Section 4.5) we find $\rho_{q1} = \gamma\rho'_{q1}$.
The 1st Maxwell equation (3.45) gives the connection between electric charge and the corresponding electric field in the rest frame O'. The 4th Maxwell Equation (3.48) gives the connection between electric currents and the corresponding magnetic fields. The discussed scenario is a stationary situation and therefore the time derivatives in the Maxwell equations vanish. Inside the charge line this results in the simplified set of Maxwell's equations,

$$\nabla \cdot \mathbf{E} = \frac{\rho_{q1}}{A\epsilon_0} \quad , \qquad \nabla \cdot \mathbf{B} = 0 \tag{3.61}$$

$$\nabla \times \mathbf{E} = 0 \quad , \qquad \nabla \times \mathbf{B} = \frac{\rho_{q1}}{Ac\epsilon_0}\beta \,. \tag{3.62}$$

A is the cross-section of the charge line, and we may get the fields outside A, where no charges and electric currents occur with the equations,

$$\nabla \cdot \mathbf{E} = 0 \quad , \qquad \nabla \cdot \mathbf{B} = 0 \tag{3.63}$$

$$\nabla \times \mathbf{E} = 0 \quad , \qquad \nabla \times \mathbf{B} = 0 \,. \tag{3.64}$$

From this, it is seen that the solutions of these equations comprise no B-fields for $\beta = 0$, and that the B-fields have azimuthal components B_φ only, while the E-fields have radial components only.

To solve these equations, we write them in cylindrical coordinates z, r, φ, where r is the distance to the z-axis and φ the transverse angle. The symmetry of the setup yields

$$\frac{\partial}{\partial z} = 0\,, \quad \frac{\partial}{\partial \varphi} = 0\,, \tag{3.65}$$

for the divergence of a vector field $\mathbf{u}$

$$\nabla \cdot \mathbf{u} = \frac{1}{r}\frac{\partial(ru_r)}{\partial r} = 0\,, \tag{3.66}$$

and for the φ-component, the only component of the curl that could be non-zero,

$$(\nabla \times \mathbf{u})_\varphi = \frac{1}{r}\frac{\partial(ru_\varphi)}{\partial r} = 0\,. \tag{3.67}$$

Eqns. (3.66) and (3.67) can be solved by the separation of variables,

$$u_r = \frac{C_1}{r} + C_2\,, \quad u_\varphi = \frac{C_3}{r} + C_4\,, \tag{3.68}$$

where C_1, C_2, C_3 and C_4 are integration constants. Physically reasonable solutions must vanish at infinity, thus $C_2 = 0$ and $C_4 = 0$. Therefore, the general solutions for Eqs. (3.63) are

$$E_z = 0\,, \quad E_r = \frac{e_1}{r}\,, \quad E_\varphi = \frac{e_3}{r}\,, \tag{3.69}$$

$$B_z = 0\,, \quad B_r = \frac{b_1}{r}\,, \quad B_\varphi = \frac{b_3}{r}\,, \tag{3.70}$$

where e_i and b_i are coefficients that must be determined with the given boundary conditions on the z-axis. Since $\nabla \times \mathbf{E} = 0$ and $\nabla \cdot \mathbf{B} = 0$ also on the z-axis, we get $E_\varphi = 0$ or $e_3 = 0$ and $B_r = 0$ or $b_1 = 0$, i.e. the electric field always points radially to or away from the z-axis and the magnetic field is tangential to concentric circles around the z-axis.

In the frame O' co-moving with the charge, $\mathbf{j}' = 0$ everywhere, thus $\nabla \times \mathbf{B}' = 0$ everywhere, and with vanishing field at infinity, $\mathbf{B}' = 0$ everywhere.

Although we have a finite linear charge density on the z-axis, the spatial charge density is infinite on the z-axis. To circumvent this singularity in the differential equation, we apply the integral of the equation over a cylindrical volume symmetric about the z-axis with height dz (Figure 3.4a),

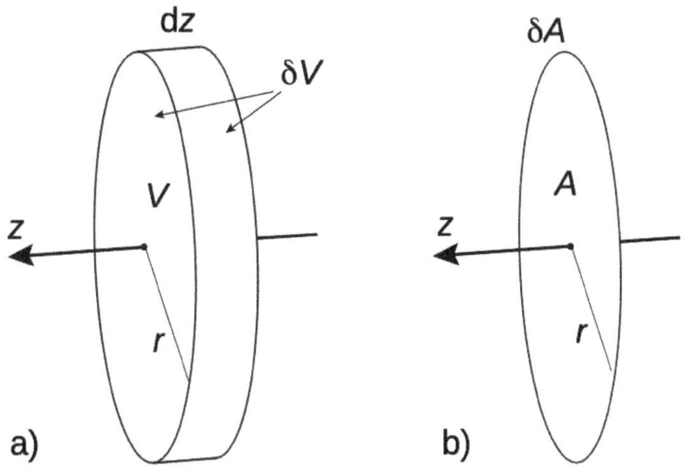

Fig. 3.4: Domain of integration: a) cylindrical volume V and surface δV symmetric to the z-axis, b) circular disc with area A and circumference δA symmetric to the z-axis.

$$\int_{V'} \nabla \cdot \mathbf{E}'\, \mathrm{d}V' = \frac{1}{\epsilon_0} \int_{V'} \rho'_{q1}\, \mathrm{d}V'\,. \tag{3.71}$$

Applying the theorem of Gauss and the cylindrical symmetry, this yields

$$\int_{\partial V'} \mathbf{E} \cdot \mathrm{d}\mathbf{O} = E'_r \, 2\pi r \, \mathrm{d}z' = \frac{\rho'_{q1}}{\epsilon_0} \, \mathrm{d}z' \,, \tag{3.72}$$

and the radial component of the electric field is

$$E'_r = \frac{\rho'_{q1}}{2\pi \epsilon_0 \, r} \,. \tag{3.73}$$

In the laboratory frame O, in which the charges move with velocity β, the electric current density is given by

$$j \equiv |\mathbf{j}| = \gamma \, \rho'_{q1} \, \beta c / A \,. \tag{3.74}$$

And the electric field strength in the laboratory frame is

$$E_r = \gamma \, E'_r = \gamma \, \frac{\rho'_{q1}}{2\pi \, \epsilon_0 \, r} \,. \tag{3.75}$$

The density of the electric current is infinite on the z-axis. To circumvent this singularity in the differential equation, we apply the integral of the equation over a circular disk concentric about the z-axis (Figure 3.4b),

$$\iint_A (\nabla \times \mathbf{B}) \, \mathrm{d}\mathbf{A} = \frac{1}{\epsilon_0 \, c^2} \iint_A \mathbf{j} \cdot \mathrm{d}\mathbf{A} = \mu_0 \, I \,, \tag{3.76}$$

where I is the total electric current through the disk (on the z-axis). Applying the theorem of Stokes of vector calculus and the cylindrical symmetry, this yields

$$\int_{\partial A} \mathbf{B} \cdot \mathrm{d}\mathbf{l} = 2\pi \, r \, B_\varphi = \frac{\gamma \, \rho'_{q1} \, \beta}{\epsilon_0 \, c} \,, \tag{3.77}$$

thus, the tangential component of the magnetic field in O is

$$B_\varphi = \frac{\rho'_{q1} \, \beta \gamma}{2\pi \, r \, \epsilon_0 \, c} = \frac{\beta}{c} E_r \,. \tag{3.78}$$

The electric and magnetic field vectors produced by a linear moving chain of charged particles, Eqs. (3.75) and (3.78), respectively, both decrease with the inverse of the distance r to the chain. While the electric field vectors point radially the magnetic field vectors point tangentially. This suggests that the effect of an infinitesimally short element of the moving charge produces a magnetic field that decreases with the inverse square of the distance to the element, similar to the electric field of a single charge, which is part of the law of Biot-Savart.

3.3.3 Coulomb and Lorentz force

Maxwell's equations state the properties of the electric and magnetic fields as a result of the occurrence of stationary and moving charges. The action of the fields back onto the charges requires an additional hypothesis that is not contained in Maxwell's equations. The forces on charges are proportional to the electric and magnetic fields at the sites of the charges.

Coulomb force

The Coulomb force law is usually stated in its original form (Coulomb, 1785a,b). The magnitude of the electric force of interaction between two (point-like) charges is proportional to the magnitudes of each of the charges and inversely proportional to the square of the distance between them. In vectorial form, this becomes

$$\mathbf{F}_{\mathrm{C}_{2,1}} = C\,\frac{q_1\,q_2}{r^2}\,\frac{\mathbf{r}}{r}\,, \tag{3.79}$$

where $\mathbf{F}_{\mathrm{C}_{2,1}}$ is the Coulomb of force of charge q_2 acting on the charge q_1, $\mathbf{r} = \mathbf{r}_2 - \mathbf{r}_1$ is the vector between the two charges q_1 and q_2 at their respective positions $\mathbf{r}_1$ and $\mathbf{r}_2$, r is the length of vector $\mathbf{r}$ and C is a proportionality constant. The force $\mathbf{F}_{\mathrm{C}_{1,2}}$ reacting on q_2 is $-\mathbf{F}_{\mathrm{C}_{2,1}}$ due to Newton's third law.

From the Gauss law and the first Maxwell equation Eq. (3.45), the proportionality constant C gets

$$C = \frac{1}{4\,\pi\,\epsilon_0}\,, \tag{3.80}$$

where the Coulomb constant C is within the given definition of $\mathbf{r}$, and the fact that charges with equal sign of repel each other, positive. For an electrostatic situation, where the charges are not moving, we get for the force on a charge q,

$$\mathbf{F}_{\mathrm{C}} = q\,\mathbf{E}\,, \tag{3.81}$$

where $\mathbf{E}$ is the electric field at the site of q. We also note that the field of q at the site of q cancels.

Lorentz force

If a charge q is moving in the electric field that is generated by all other charges, q experiences an additional force proportional to the magnetic field, which is a consequence of the Lorentz transformation of the electric field (see Section 3.3.2). We find the Lorentz force acting on q:

$$\mathbf{F}_{\mathrm{L}} = q\,(\mathbf{E} + \mathbf{v} \times \mathbf{B})\,, \tag{3.82}$$

where q is moving with velocity $\mathbf{v}$ in the electric and magnetic field $\mathbf{E}$ and $\mathbf{B}$. In some text books the $q\,\mathbf{E}$ term is neglected, though we stick here with the general formulation that includes the Coulomb force in the Lorentz force, also because **E** and **B** fields depend on the frame of reference. Notably, the force due to the **B** field acts perpendicular to the velocity, which can be understood on the base of the above Section 3.3.2. The handedness of the force due to the magnetic field is imposed by the law of Ampère Eq. (3.48).

4 Special Relativity Revisited

A new theory is difficult to be accepted if one tries to understand its consequences within the framework of the "old" or existing theory, especially if the new theory includes a shift in paradigm. In this Chapter we discuss misconceptions and so-called paradoxa such as that of the ether drift experiment or the space travelling twins. Furthermore, the observability of the one-way velocity of light and of time dilation and length contraction are reviewed.

4.1 Ether drift experiments

Any mechanical wave relies on a carrier medium as it is air for sound waves, or water for water waves. From this it was postulated that light waves rely as well on such a carrier medium or "ether". As sound waves allow e.g. the detection of air movements like wind, it was obvious to think of experiments with light waves that would indicate drifts or movements of the light bearing ether. For this purpose the movement of the Earth around the sun was the closest velocity to that of the light and available for experiments. Although we have seen in Section 3.3 that certain solutions of the Maxwell equations allow light to propagate in vacuum, in this Section we will revisit experiments that searched for an ether.

Textbooks on special relativity sometimes present the Michelson-Morley experiment extensively, firstly as the *experimentum crucis* that was decisive for Einstein's work and secondly, as an ether drift experiment that disproved the existence of the light-bearing ether. Both emphases are only partly correct. In this Section the aspect of the ether drift experiment is discussed in some detail.

In fact, the Michelson-Morley experiment (see Section 1.3.2) does not disprove the existence of an ether and its corresponding absolute system of reference (Shupe, 1985). The interesting point is that the experiments are in contradiction with an absolute frame defined by an ether in the context of classical mechanics, which cannot explain the *null result* for a finite speed of light. In the context of special relativity the experiments lose their conclusiveness to access an ether drift. To illustrate this, the operational concept of length and time measurements must be tested. The Michelson-Morley experiment is a two way experiment where two reflected light beams interfere, i.e. where spatially and temporally coinciding events are observed. Separate events require synchronization as discussed in Section 1.3.2. It does e.g. exclude the length measurement that includes the instantaneous reading of markings on a rod. Generally, one-way experiments are considered impossible (Ahrens, 1962). In Sections 4.4 and 4.5 the non-obvious consequences of a possible anisotropy of the speed of light will be further analyzed.

Here we investigate whether an ether may be observed within special relativity in a picture with a constant speed of light with respect to an ether. The considerations are

https://doi.org/10.1515/9783111503592-005

based on an additional system of hypotheses for the transformation between systems of reference:

1. An ether, evenly filling space, defines an absolute frame of reference.
2. The ether carries electromagnetic waves with the characteristic velocity of propagation c.

This velocity may serve as a standard for length and time measurements. In the frame of the ether, we define a clock based on two transmitter-receivers A and B which are separated by the distance Δs (Shupe, 1985). The emitter sends a pulse at the moment it receives a pulse. In the rest frame of the ether, the transition time of a light pulse from A to B or reverse is

$$\Delta t = \frac{\Delta s}{c}\,. \tag{4.1}$$

If the clock moves in a direction normal to the axis AB with the velocity $v=\beta\, c$, the transition time from A to B measured in the ether frame becomes

$$\Delta t' = \frac{\Delta t}{\sqrt{1-\beta^2}} = \gamma\,\Delta t\,. \tag{4.2}$$

The *null result* of ether drift experiments requires isotropy, i.e. the transition time must be the same $\Delta t'$ as in Eq. (4.2) independent on the direction of motion of the clock. If the clock moves parallel to the axis AB, this can only be achieved if the distance $\overline{AB}$ measured in the ether frame is

$$\Delta s' = \frac{\Delta s}{\gamma}\,. \tag{4.3}$$

The *time dilation* of Eq. (4.2) and the *length contraction* of Eq. (4.3) require, that the space-time transformation between the ether frame and the moving frame of the clock is a Lorentz transformation (Section 2.3). Lorentz transformations form an algebraic group, i.e. the composition of Lorentz transformations forms Lorentz transformations. Thus, two frames that move uniformly with respect to the ether frame are also connected with a Lorentz transformation with the same characteristic velocity c, and thus, operationally not distinguishable from the ether frame.

To emphasize this, we look at the relativistic acoustic Doppler effect, see Bachman (1982). Acoustic waves travel about 6 orders of magnitude slower than light, therefore, for all purposes, it can be treated in classical form. Bachman (1982) presented the relativistic acoustic Doppler effect for some "persistent puristic" reasons. We just give the result here (for a detailed derivation please consult Bachman (1982)),

$$\frac{\nu'}{\nu_0} = \frac{u+v_0}{u-v_s}\sqrt{\frac{1-\frac{v_s^2}{c^2}}{1-\frac{v_0^2}{c^2}}}\,, \tag{4.4}$$

where ν' is the frequency of the waves received by an observer, ν_0 is the proper frequency of the source, u the wave velocity relative to the wave-carrying medium, and the relativistic limit velocity is c. The velocity of the source toward the observer relative to the medium is v_s and the velocity of the observer toward the source relative to the medium is v_0. The relativistic composition of v_s and v_0

$$v_r = \frac{v_s + v_0}{1 + \frac{v_s\, v_0}{c^2}} \tag{4.5}$$

is the velocity of the observer relative to the source.

To recover the equation for the Doppler shift for light, we substitute c for the wave velocity u in Eq. (4.4),

$$\frac{\nu'}{\nu_0} = \frac{c + v_0}{c - v_s} \sqrt{\frac{1 - \frac{v_s^2}{c^2}}{1 - \frac{v_0^2}{c^2}}}. \tag{4.6}$$

With a few mathematical operations, we obtain

$$\left(\frac{\nu'}{\nu_0}\right)^2 = \frac{(c + v_0)^2}{(c - v_s)^2} \cdot \frac{c^2 - v_s^2}{c^2 - v_0^2} = \frac{(c + v_0)\,(c + v_s)}{(c - v_s)\,(c - v_0)} = \frac{1 + c\frac{v_0 + v_s}{c^2 + v_0\, v_s}}{1 - c\frac{v_0 + v_s}{c^2 + v_0\, v_s}}, \tag{4.7}$$

and with the velocity composition law, Eq. (4.5),

$$\frac{\nu'}{\nu_0} = \sqrt{\frac{1 + \frac{v_r}{c}}{1 - \frac{v_r}{c}}}, \tag{4.8}$$

and thus, we recovered the relativistic Doppler shift formula for light waves. It is interesting to note that the wave carrying medium does not play any role in this equation, neither v_s nor v_0, thus *The ether just fades away* (Mirabelli, 1985).

In this sense, the ether frame is not in contradiction with so called ether drift experiments, such as the experiment of Michelson and Morley (1887). The ether frame is not distinguishable from other frames, and thus not observable. It is therefore not inconsistent with special relativity, but has no physical significance, and is therefore not necessary:

> The choice is traditionally made to accept the postulates of special relativity (*the relativity principle and the isotropy of the speed of light). But it should be emphasized that though the postulates of special relativity imply the Lorentz transformations, the reverse is not true (Shupe, 1985)*

4.2 Superluminal motion

Several articles of astrophysical measurements of expanding relativistic jets from quasars reported *superluminal expansion velocities* (Pearson et al., 1981). At first glance, this seems to contradict special relativity. Closer scrutiny reveals confusion and misuse of terminology.

The reported example is the relativistic jet of quasar 3C273 at a distance of $2.4 \cdot 10^9$ ly (light-years), of which the measured apparent angular distance between the center and a gas cloud increases by 0.76±0.04 parsec per year. This corresponds to expansion speed of 8.8 times the speed of light.

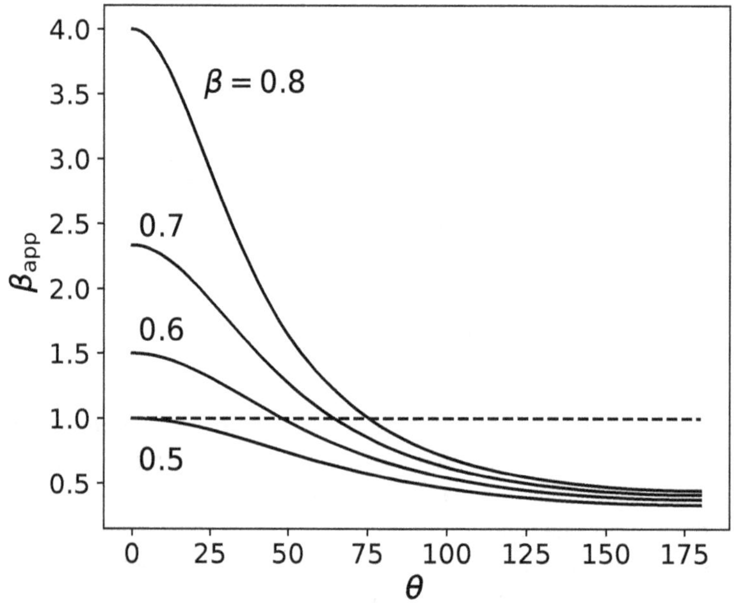

Fig. 4.1: Apparent speed β_{app} of a point in space as supposedly observed with the expanding gas jets of a quasar.

We analyze the motion of the gas cloud by the motion of a luminous point P moving with velocity v away from the pulsar. The direction of the motion includes an angle θ with the direction from the pulsar to an observer (e.g. Earth) at a distance d from the pulsar. The distance of P to the pulsar at a given time t_p is

$$s_p = v\, t_p \,. \tag{4.9}$$

The light (e.g. a pulse) emitted at the time t_{P} at P will be observed on Earth at time

$$t_{\mathrm{B}} = t_{\mathrm{P}} + \frac{1}{c}\left(\frac{d}{\sin\theta} + v\,t_{\mathrm{P}}\cos\theta\right) \tag{4.10}$$

Substituting Eq. (4.9) in Eq. (4.10) and differentiate

$$\frac{\mathrm{d}t_B}{\mathrm{d}z_p} = \frac{1}{v} - \frac{1}{c}\cos\theta\,, \tag{4.11}$$

which is the inverse of what at best can be called the *apparent velocity* v_{app} of the point as seen from Earth,

$$v_{\mathrm{app}} = \frac{\mathrm{d}z_{\mathrm{P}}}{\mathrm{d}t_{\mathrm{B}}} = \frac{v}{c - v\cos\theta}\,c = \frac{\beta}{1-\beta\cos\theta}\,c\,. \tag{4.12}$$

For high enough velocities and if the point is approaching the observer, v_{app} may be larger than the speed of light c (Figure 4.1).

This is the result of a flawed time measurement. If the point emits pulses at regular intervals in the frame of the point, the observer receives these pulses at intervals that are Doppler shifted. Thus, the apparent velocity v_{app} does not correspond to the *coordinate velocity*, which cannot exceed the speed of light. In Section 3.1, emphasis is put on accurate definitions of different measures of the rate of change in position, where the corresponding spatial and the temporal intervals may be measured in different frames of reference. A careful analysis of the qualities of these measures may prevent similar confusions as stated above.

4.3 Relativistic mass

Concepts of mass were mentioned in Section 1.4 as hypotheses beyond those for bare space and time. Still, moving masses are described in the framework of the special theory of relativity, where often the misleading term *relativistic mass* is found.

In 1948 Einstein wrote in a letter to Lincoln Barnett, the author of the book "The Universe and Dr. Einstein":

> Es ist nicht gut von der Masse $M = m/\sqrt{1-v^2/c^2}$ eines bewegten Körpers zu sprechen, da für M keine klare Definition gegeben werden kann. Man beschränkt sich besser auf die Ruhe-Masse m.
> *It is not good to introduce the concept of the mass $M = m/\sqrt{1-v^2/c^2}$ of a moving body for which no clear definition can be given. It is better to introduce no other mass concept than the "rest mass" m.*
> From Okun (1989).

In the quote M is the so called "relativistic mass", a term that can be avoided as we saw in Section 3.2 on relativistic dynamics.

The derivations of the relativistic second law and the relativistic centripetal force for a circular motion use the viewpoint of the accelerated observer (see Section 3.2.1).

The relativistic laws of the action of forces on masses result primarily from kinematic arguments. The only dynamic argument included is the fact that accelerated frames can be distinguished from non-accelerated frames by the occurrence of a force within the accelerated frame with no reference to any specific frame of reference (see Section 3.1.2). The product $\gamma\, m$ in Eq. (3.23), often referred to as *relativistic mass*, only occurs if a specific frame of reference is involved. In this work, we do not introduce and use the term relativistic mass and stick to what we call *mass* or rest mass, which is an invariant. This can be considered merely as a “radical” or “aesthetic” argument against the use of the term “relativistic mass”.

4.4 Isotropy and the one-way velocity of light

Einstein (1905a) recognized the need of a "definition" for the synchronization of separate clocks:

> Befindet sich im Punkte A des Raumes eine Uhr, so kann ein in A befindlicher Beobachter die Ereignisse in der unmittelbaren Umgebung von A zeitlich werten durch Aufsuchen der mit diesen Ereignissen gleichzeitigen Uhrzeigerstellungen. Befindet sich auch im Punkte B des Raumes eine Uhr - wir wollen hinzufügen, “eine Uhr von genau derselben Beschaffenheit wie die in A befindliche”- so ist auch eine zeitliche Wertung der Ereignisse in der unmittelbaren Umgebung von B durch einen in B befindlichen Beobachter möglich. Es ist aber ohne weitere Festsetzung nicht möglich, ein Ereignis in A mit einem Ereignis in B zeitlich zu vergleichen; wir haben bisher nur eine „A-Zeit“ und eine „B-Zeit“, aber keine für A und B gemeinsame “Zeit” definiert. Die letztere Zeit kann nun definiert werden, indem man *durch Definition* festsetzt, daß die “Zeit”, welche das Licht braucht, um von A nach B zu gelangen, gleich ist der „Zeit“, welche es braucht, um von B nach A zu gelangen. Es gehe nämlich ein Lichtstrahl zur “A-Zeit” t_A von A nach B ab, werde zur „B-Zeit“ t_B in B gegen A zu reflektiert und gelange zur “A-Zeit” t'_A nach A zurück. Die beiden Uhren laufen definitionsgemäss synchron, wenn
>
> $$t_B - t_A = t'_A - t_B\,. \tag{4.13}$$
>
> Wir nehmen an, daß diese Definition des Synchronismus in widerspruchsfreier Weise möglich sei, und zwar für beliebig viele Punkte, daß also allgemein die Beziehungen gelten:
>
> 1. Wenn die Uhr in B synchron mit der Uhr in A läuft, so läuft die Uhr in A synchron mit der Uhr in B.
>
> 2. Wenn die Uhr in A sowohl mit der Uhr in B als auch mit der Uhr in C synchron läuft, so laufen auch die Uhren in B und C synchron relativ zueinander.
>
> Wir haben so unter Zuhilfenahme gewisser (gedachter) physikalischer Erfahrungen festgelegt, was unter synchron laufenden, an verschiedenen Orten befindlichen, ruhenden Uhren zu verstehen ist und damit offenbar eine Definition von “gleichzeitig” und „Zeit“ gewonnen. Die “Zeit” eines Ereignisses ist die mit dem Ereignis gleichzeitige Angabe einer am Orte des Ereignisses befindlichen, ruhenden Uhr, welche mit einer bestimmten, ruhenden Uhr, und zwar für alle Zeitbestimmungen mit der nämlichen Uhr, synchron läuft.
>
> Wir setzen noch der Erfahrung gemäß fest, daß die Größe
>
> $$\frac{2AB}{t'_{\mathrm{A}} - t_{\mathrm{A}}} = V\,, \tag{4.14}$$
>
> eine universelle Konstante (die Lichtgeschwindigkeit im leeren Raume) sei.

Wesentlich ist, daß wir die Zeit mittels im ruhende System ruhender Uhren definiert haben; wir nennen die eben definierte Zeit wegen dieser Zugehörigkeit zum ruhenden System „die Zeit des ruhenden Systems“

If at the point A of space there is a clock, an observer at A can determine the time values of events in the immediate proximity of A by finding the positions of the hands which are simultaneous with these events. If there is at the point B of space another clock in all respects resembling the one at A, it is possible for an observer at B to determine the time values of events in the immediate neighbourhood of B. But it is not possible without further assumption to compare, in respect of time, an event at A with an event at B. We have so far defined only an "A time" and a "B time". We have not defined a common "time" for A and B, for the latter cannot be defined at all unless we establish by definition that the "time" required by light to travel from A to B equals the "time" it requires to travel from B to A. Let a ray of light start at the "A time" t_A *from A towards B, let it at the "B time"* t_B *be reflected at B in the direction of A, and arrive again at A at the "A time"* t'_A.

In accordance with definition the two clocks synchronize if

$$t_B - t_A = t'_A - t_B \,. \tag{4.15}$$

We assume that this definition of synchronism is free from contradictions, and possible for any number of points; and that the following relations are universally valid:—

1. If the clock at B synchronizes with the clock at A, the clock at A synchronizes with the clock at B.

2. If the clock at A synchronizes with the clock at B and also with the clock at C, the clocks at B and C also synchronize with each other.

Thus with the help of certain imaginary physical experiments we have settled what is to be understood by synchronous stationary clocks located at different places, and have evidently obtained a definition of "simultaneous", or "synchronous", and of "time". The "time" of an event is that which is given simultaneously with the event by a stationary clock located at the place of the event, this clock being synchronous, and indeed synchronous for all time determinations, with a specified stationary clock.

In agreement with experience we further assume the quantity

$$\frac{2AB}{t'_A - t_A} = c\,, \tag{4.16}$$

to be a universal constant - the speed of light in empty space.

It is essential to have time defined by means of stationary clocks in the stationary system and the time now defined being appropriate to the stationary system we call "the time of the stationary system". from Einstein (1905a), English translation from "The Principle of Relativity", 1923, Methuen and Company, Ltd., London.

The convention of the synchronization has fundamental consequences for the very basic qualities assumed for space. Seemingly, all relativistic experiments, both consistent thought experiments and real measurements, were recognized as two-way experiments. This means that e.g. transition times for light beams are always measured with synchronized clocks according to Einstein (1905a) or for round trips of light beams with one clock at the same starting and end point. All known two-way experiments show an isotropic two-way velocity of light. As an example we think of a light spark that is emitted in the center of a sphere with a mirror surface. It is found that the reflected light arrives from all directions at the same point and time from where it was emitted.

The controversy on the observability of the isotropy of the one-way velocity of light could not been resolved so far. Many papers present thought experiments to measure the one-way velocity of light, but were refuted by papers that showed a flaw in the conception, e.g. Weinberger and Mossel (1971) versus Stedman (1972), Stedmann (1973), Gordon (1972) and Ehrlichson (1973), or Feenberg (1974) versus Øhrstrom (1980), or Brehme (1988) versus Ungar (1988), or Fung and Hsieh (1980) versus Nissim-Sabat (1982), or Coleman and Korté (1992) versus Anderson and Stedman (1992, 1994) and Greaves et al. (2009) versus Finkelstein (2010), which could not resolve their differences.

We do not go into the details of these debates, yet, it is worth mentioning the formulation of a larger class of transformations, the so called ϵ-transformations, of which the Lorentz-transformation is a special case (Ungar, 1986, 1991b). The anisotropy of the one-way velocity of light is accounted for by an anisotropy parameter $-1 < \epsilon < 1$ that takes the value of $\epsilon = 0$ for the standard (isotropic) Lorentz transformation. Two-way experiments can be interpreted with an ϵ-transformation, and also with the assumption of an anisotropic one-way velocity of light. The results for the two-way experiments are identical with the assumption of an isotropic two-way speed of light, thus the ϵ-transformations provide an argument against the one-way isotropy assumption (Anderson and Stedman, 1992, 1994; Vetharaniam and Stedman, 1991). If this proves to be true, the isotropy of the one-way velocity of light must be accepted as being not observable. It is astonishing that such a fundamental property of spacetime may not be observed. This fact does not reduce the value of special relativity since all known observations are also consistent with the value $\epsilon = 0$ of the Lorentz transformation. It is thus allowed to apply Ockham's razor and to choose the simplest possible transformation.

In the following, discuss a scenario for the description of the one-way velocity of light and use the notation of Ungar (1991b). All quantities, such as velocity, time, and space will be written in capital letters if they refer to the anisotropic transformation.

To determine the simultaneity of events at two spatially distant locations A and B, one needs to have two clocks and a procedure to synchronize them. Let a light signal be emitted from point A at time T_1 and reflected back at point B at time T_2 to return at point A at time T_3,

$$T_2 - T_1 = \epsilon_R \left(T_3 - T_1\right), \tag{4.17}$$

with the Reichenbach anisotropy parameter ϵ_R, $(0 < \epsilon_R < 1)$. Then, according to Reichenbach (1969) and Ungar (1991b), the clock at point B at time T_2 is said to be synchronous with the clock at Point A. The parameter ϵ_R is a synchrony parameter and its restriction ensures the validity of the causality condition. Now the one-way velocity of light in the positive direction (from A to B), C_+, can be determined with the two-way speed of light c,

$$C_+ = \frac{d}{T_2 - T_1} = \frac{1}{2\,\epsilon_R}\,\frac{2d}{T_3 - T_1} = \frac{c}{2\,\epsilon_R}\,, \tag{4.18}$$

where d is the distance between A and B. Furthermore, the sum of the two time intervals from A to B and back from B to A gives

$$(T_2 - T_1) + (T_3 - T_2) = \frac{d}{C_+} + \frac{d}{-C_-} \equiv \frac{2d}{c}\,. \tag{4.19}$$

Thus, the inverse of the two-way speed of light is the harmonic mean of the one-way velocities C_+ and C_-,

$$\frac{1}{c} = \frac{1}{2}\left(\frac{1}{C_+} + \frac{1}{-C_-}\right). \tag{4.20}$$

For further use (Ungar, 1991b), we redefine the synchrony or anisotropy parameter,

$$\epsilon = 1 - 2\,\epsilon_R\,, \quad (0 < \epsilon < 1) \tag{4.21}$$

and rewrite the one-way velocities with the two-way velocity c and the new anisotropy parameter ϵ,

$$C_+ = \frac{c}{1-\epsilon} \quad \text{and} \quad C_- = \frac{-c}{1+\epsilon}\,. \tag{4.22}$$

The anisotropic Lorentz transformation $\mathbf{L}(V)$ with the velocity V is

$$Z' = \Gamma\,[(1 + p\,V)\,Z + V\,T] \quad , \quad T' = \Gamma\,(T + q^2\,V\,Z)\,, \tag{4.23}$$

where the parameters p, q^2 and Γ^2 are given by

$$-\,p = \frac{1}{C_+} + \frac{1}{C_-} = -\frac{2\,\epsilon}{c}\,, \tag{4.24}$$

$$-\,q^2 = \frac{1}{C_+\,C_-} = -\frac{1-\epsilon^2}{c^2}\,, \tag{4.25}$$

and

$$\Gamma^{-2} = \left(1 - \frac{V}{C_+}\right)\left(1 - \frac{V}{C_-}\right). \tag{4.26}$$

Written in matrix notation this is

$$\begin{pmatrix} T' \\ Z' \end{pmatrix} = \mathbf{L}\cdot\begin{pmatrix} T \\ Z \end{pmatrix} = \Gamma\begin{pmatrix} 1 & q\,V \\ q\,V & 1 + p\,V \end{pmatrix}\cdot\begin{pmatrix} T \\ Z \end{pmatrix}. \tag{4.27}$$

These transformations constitute a group where the composition of transformations

$$\Gamma_{12}\begin{pmatrix} 1 & q\,V_{12} \\ q\,V_{12} & 1 + p\,V_{12} \end{pmatrix} =$$

$$\Gamma_1\begin{pmatrix} 1 & q\,V_1 \\ q\,V_1 & 1 + p\,V_1 \end{pmatrix}\cdot\Gamma_2\begin{pmatrix} 1 & q\,V_2 \\ q\,V_2 & 1 + p\,V_2 \end{pmatrix} \tag{4.28}$$

constitutes 4 equations for Γ_{12},Γ_1, Γ_2, V_{12}, V_1 and V_2,

$$\Gamma_{12}\begin{pmatrix} 1 & q\,V_{12} \\ V_{12} & 1+p\,V_{12}\end{pmatrix} =$$

$$\Gamma_1\,\Gamma_2 \quad \begin{pmatrix} 1+q^2\,V_1\,V_2 & q\,V_2+q\,V_1\,(1+p\,V_2) \\ q\,V_1+q\,V_2\,(1+p\,V_1) & q^2\,V_1\,V_2+(1+p\,V_1)\,(1+p\,V_2)\end{pmatrix} . \tag{4.29}$$

Eliminating Γ_1, Γ_2 and Γ_{12} yields the velocity composition law

$$V_{12} = \frac{V_1+V_2+p\,V_1V_2}{1+q^2\,V_1V_2} . \tag{4.30}$$

The composition of a transformation $\mathbf{L}(V)$ with its inverse transformation gives

$$\mathbf{L}(V_+)\cdot\mathbf{L}^{-1}(V_-) = E = L(0)\,, \tag{4.31}$$

and therefore the nominator of Eq. (4.30) must vanish,

$$V_+ + V_- + p\,V_+\,V_- = 0\,, \tag{4.32}$$

thus,

$$V_+ = \frac{-V_-}{1-p\,V_-} = \frac{-V_-}{1-\dfrac{2\epsilon}{c}\,V_-} \quad , \quad V_- = \frac{-V_+}{1+p\,V_+} = \frac{-V_+}{1+\dfrac{2\epsilon}{c}\,V_+} \tag{4.33}$$

This is the *anisotropic one-way velocity reciprocity principle*, i.e. V_+ is the velocity of the coordinate frame B with respect to frame A, and V_- is the velocity of frame A with respect to frame B. Eliminating V_+ and V_- with Eq. (4.30) yields V_+ and V_- written with the two-way velocity v and the two-way velocity of light c,

$$V_+ = \frac{v}{1+\epsilon\,\dfrac{v}{c}} \quad , \quad V_- = \frac{-v}{1+\epsilon\,\dfrac{v}{c}} . \tag{4.34}$$

From Eqs. (4.34) we also obtain

$$\frac{1}{v} = \frac{1}{2}\left(\frac{1}{V_+} - \frac{1}{V_-}\right), \tag{4.35}$$

i.e. the two-way velocity v is the harmonic mean of V_+ and V_-, as the two-way speed of light c is the harmonic mean of C_+ and C_-, see Eq. (4.20).

We illustrate the unobservability of the one-way velocities with the longitudinal Doppler shift. An electromagnetic wave with wavelength λ and frequency ν moves along the Z-axis of frame F in the direction of the positive Z-axis with velocity C_+, thus $\lambda\nu = C_+$. An observer O' in the origin of frame F' moves with the velocity V_+ relative to frame F. He coincides with the observer O at the origin of frame F at the moment, when a wave crest passes the origins. Since observer O' stays at the origin of F' ,

$$\Delta Z' = 0\,. \tag{4.36}$$

The time interval between the crossing of the wave crest and the following wave crest can be expressed with the frequency ν

$$\Delta T' = \frac{1}{\nu'} . \tag{4.37}$$

The distance which the observer O' moves in the frame F is

$$\Delta Z = V_+ \Delta T . \tag{4.38}$$

Since the wave velocity is C_+, the time interval in Eq. (4.37) now measured in F is

$$\Delta T = \frac{\lambda + \Delta Z}{C_+} = \frac{1}{\nu} + \frac{V_+}{C_+} \Delta T , \tag{4.39}$$

and solving this equation for electromagnetic waves yields the connection between ΔT and the frequency ν

$$\Delta T = \frac{1}{\nu \left(1 - \frac{V_+}{C_+}\right)} \tag{4.40}$$

Applying the anisotropic Lorentz transformation, Eq. (4.23), with $\Delta Z' = 0$, yields

$$\frac{1}{\nu' \sqrt{\left(1 - \frac{V_+}{C_+}\right)\left(1 - \frac{V_+}{C_-}\right)}} = \frac{1}{\nu \left(1 - \frac{V_+}{C_+}\right)} , \tag{4.41}$$

and finally the equation for the anisotropic longitudinal Doppler shift,

$$\frac{\nu}{\nu'} = \sqrt{\frac{1 - \frac{V_+}{C_-}}{1 - \frac{V_+}{C_+}}} . \tag{4.42}$$

Substituting C_+ and V_+ from Eqs. (4.22) and (4.34) recovers the equation of the standard (isotropic) special relativity theory,

$$\frac{1 - \frac{V_+}{C_-}}{1 - \frac{V_+}{C_+}} = \frac{1 + \frac{v}{c}}{1 - \frac{v}{c}} , \tag{4.43}$$

with the coordinate velocity $\beta = v/c$. Similarly, the same results can be obtained for all combinations of the velocities V_+ and V_- with the velocities of light, C_+ and C_-.

This is an interesting result. If the one-way velocities cannot be unambiguously measured, the velocities c and v must always be interpreted as two-way velocities. The anisotropic (ϵ-) Lorentz transformation also offers an objective criterion for observable or unobservable quantities: if a term for a quantity contains ϵ, then this quantity is not observable. For instance, the anisotropic one-way reciprocity principle in Eq. (4.33) has no physical significance because it is not observable. Similarly, the time dilation and length contraction are not observable (see Section 4.5), and thus, the twin paradox has no physical significance. Someday perhaps, someone finds a way to unambiguously determine (measure) the one-way velocity of light and with this, the anisotropy parameter ϵ.

4.5 Length and time

We assume the scenario as defined in Figure 2.5 where the frame O' moves with velocity β in the direction of the positive z-axis of frame O. An event (t', z') in frame O' is transformed back to frame O by the inverse Lorentz transformation,

$$\begin{pmatrix} c\,t \\ z \end{pmatrix} = \begin{pmatrix} \gamma & \beta\gamma \\ \beta\gamma & \gamma \end{pmatrix} \cdot \begin{pmatrix} c\,t' \\ z' \end{pmatrix} , \quad \begin{pmatrix} c\,t' \\ z' \end{pmatrix} = \begin{pmatrix} \gamma & -\beta\gamma \\ -\beta\gamma & \gamma \end{pmatrix} \cdot \begin{pmatrix} c\,t \\ z \end{pmatrix} . \tag{4.44}$$

Let one clock be at rest with respect to frame O', at the position $z'_1 = z'_2 = z'_c$. At two different times t'_1 and t'_2 the clock is read in frame O, according to the left equation of Eq. (4.44), which gives

$$c\,t_1 = \gamma\,c\,t'_1 + \beta\gamma\,z'_c\,, \quad c\,t_2 = \gamma\,c\,t'_2 + \beta\gamma\,z'_c\,. \tag{4.45}$$

The corresponding time intervals $\Delta t = t_2 - t_1$ and $\Delta t' = t'_2 - t'_1$ in the respective frames are related by

$$\Delta t = \gamma\,\Delta t'\,. \tag{4.46}$$

This situation is reciprocal for the frames. The clock can be held as well at a fixed position in frame O and be read in frame O', which means that in the transformation β is replaced by $-\beta$,

$$\Delta t' = \gamma\,\Delta t\,. \tag{4.47}$$

Let a rod of length $\Delta z' = z'_2 - z'_1$ be at rest with the frame O'. At equal time in O, $t_1 = t_2 = t_r$, the length $\Delta z = z_2 - z_1$ of the rod is measured in frame O by measuring the corresponding coordinates z_1 and z_2, which according to right equation of Eq. (4.44) are

$$z'_1 = -\beta\gamma\,c\,t_r + \gamma\,z_1\,, \quad z'_2 = -\beta\gamma\,c\,t_r + \gamma\,z_2\,, \tag{4.48}$$

and thus

$$\Delta z = \frac{\Delta z'}{\gamma}\,, \tag{4.49}$$

holds, again together with the reciprocal relation

$$\Delta z' = \frac{\Delta z}{\gamma}\,, \tag{4.50}$$

Eqs. (4.46), (4.47), (4.49) and (4.50) are the equations for *time dilation* and *length contraction*. They are the origin of so-called paradoxa. The history of teaching relativity accounts for many seeming paradoxa and exemplifies how faulty statements cause confusion. A paradox is a contradiction to an existing and accepted theory or doctrine, as the word "paradox" (against the doctrine) says. This is different from a riddle or an unsolved problem within a given theory, and it is distinct from a misunderstanding of the theory. The resolution of a paradox requires a different intellectual process than the solution of a riddle or a problem (Harrison, 1984).

The fact is that time dilation and length contraction are not directly observable, as can be shown by the anisotropic Lorentz transformation in Section 4.4. Substituting V_+ from Eq. (4.34), C_+ and C_- from Eqs. (4.22) in Eq. (4.26) for the factor Γ, we obtain after some algebraic manipulations,

$$\Gamma = \frac{1 - \dfrac{\epsilon\, v}{c}}{\sqrt{1 - \dfrac{v^2}{c^2}}} \,. \tag{4.51}$$

The product of the anisotropic one-way velocity with Γ

$$V_+\, \Gamma = \frac{v}{\sqrt{1 - \dfrac{v^2}{c^2}}} = v\, \gamma \tag{4.52}$$

can now be written in terms of the two-way velocity and the isotropic γ and is therefore observable. The anisotropic time dilation,

$$\Delta T = \Gamma\, \Delta T' \tag{4.53}$$

can be derived from the anisotropic Lorentz transformation, Eq. (4.23), in the same way as for the isotropic time dilation, Eq. (4.47). The anisotropic time dilation, Eq. (4.53), contains ϵ and is thus not observable. The distance d that the particle travels in the time ΔT with velocity V_+ is

$$d = V_+\, \Delta T = V_+\, \Gamma\, \Delta T' = v\, \gamma\, \Delta T' \,. \tag{4.54}$$

Since $d = \text{const}$, $\Delta T' = \Delta t'$ is independent of ϵ, thus

$$d = v\, \gamma\, \Delta t' \,, \tag{4.55}$$

which contains measurable quantities, such as the isotropic time dilation $\gamma \Delta t'$ and the velocity v, while the anisotropic time dilation $\gamma \Delta T'$ contains the unknown element ϵ that is not observable (Ungar, 1991b).

4.6 The spacetravelling twins

The so-called "twin paradoxon", is commonly assigned to the prediction that a relativistically traveling twin after return should appear younger than his twin sibling that was not traveling. Within the special theory of relativity, this is no paradoxon since the plot does not impose a symmetric history of the two siblings: the effect of different aging is consistently explained by different *acceleration* histories of the two twins. Notably it is not sufficient to quote Eqs. (4.46) and (4.47) and to claim that both twins feel the same time dilation. As we will see, at least 3 different inertial frames are involved, and this breaks the ageing symmetry between the twins.

In Section 3.1.2, spaceflight is discussed from the standpoint of astronauts, i.e. from local observers alone. The time that passed during a journey can only be compared to the time elapsed on Earth at the time of return. The communication between the twins shall be restricted to the exchange of time signals. Both twins have identical devices with clock, transmitter and receiver that register and send out signals at regular time intervals with the same proper frequency ν.

A scenario with uniform velocities is assumed, and we have to distinguish 3 observers O, O' as defined in Figure 2.5 and O". The first twin is observer O, while the travelling twin changes in the middle of the trip from O' to O".

- The first twin O sits at rest on Earth and starts his clock at the moment, when the travelling twin O' passes Earth.
- In the coordinate system of O the travelling twin O' flies with relativistic velocity β and starts her clock at the moment of coincidence with the first twin at O.
- In the coordinate system of O, O" moves with velocity $-\beta$.
- In the middle of the trip the travelling twin transforms from O' to O" and travels back to Earth.
- At the second encounter the clocks of both twins are stopped and compared.

In the coordinate system of O the transformation of observers O' into O" shall happen at distance d. The flight time Δt_d in the frame of reference of Earth is

$$\Delta t_d = \frac{d}{c\beta}, \tag{4.56}$$

and the time on Earth at the coincidence of observer O" and O is $t_2 = 2\,\Delta t_d$. The device sends N signals in the time Δt_d with the frequency

$$\nu = \frac{N}{\Delta t_d}. \tag{4.57}$$

Firstly, we interpret the observations made by observers O' and O" in comparison with the emitted signals of observer O. Observer O' measures a proper time $\Delta\tau_d$ between the coincidence with O and the coincidence with O", and O" measures the same proper time between his corresponding coincidences. The observers O' and O" register N' and N'' signals during their corresponding flight intervals, thus they observe the signals at corresponding frequencies

$$\nu' = \frac{N'}{\Delta\tau_d}, \quad \nu'' = \frac{N''}{\Delta\tau_d}. \tag{4.58}$$

The observed shifts in the frequency corresponds to the Doppler shift, Eq. (2.24),

$$\frac{N'}{\Delta\tau_d} = \frac{N}{\Delta t_d}\sqrt{\frac{1-\beta}{1+\beta}}\,, \quad \frac{N''}{\Delta\tau_d} = \frac{N}{\Delta t_d}\sqrt{\frac{1+\beta}{1-\beta}}\,. \tag{4.59}$$

Adding these two equations yields

$$\frac{N'+N''}{\Delta\tau_d} = \frac{N}{\Delta t_d}\left(\sqrt{\frac{1+\beta}{1-\beta}} + \sqrt{\frac{1-\beta}{1+\beta}}\right) = 2\,\gamma\,\frac{N}{\Delta t_d}\,, \tag{4.60}$$

and since the observers O' and O" collect all signals emitted by observer O, thus $N' + N'' = N$, we obtain with some algebraic rearrangements

$$\Delta t_d = 2\,\gamma\,\Delta\tau_d\,. \tag{4.61}$$

Secondly, we interpret vice versa the observations of observer O in comparison with the emitted signals of observers O' and O". Both observers, O' and O", send M signals during their equal proper time intervals $\Delta\tau_d$ between their corresponding coincidences. Observer O receives a total of $2M$ signals during the time interval between the coincidences with O' and O". The time of the first coincidence with O' is $t_0 = 0$ and the last signal from O' arrives at $t_1 = \Delta t_d + d/c$, thus the total time interval in which he receives M signals from O' is $\Delta t_d + d/c$. The time of the coincidence of O with O" is $t_2 = 2\,\Delta t_d$, thus the total time interval in which he receives M signals from O" is $\Delta t_d - d/c$. Again, he receives the signals Doppler shifted,

$$\nu' = \frac{M}{\Delta t_d + d/c} = \frac{M}{\Delta\tau_d}\sqrt{\frac{1-\beta}{1+\beta}}\,, \quad \nu'' = \frac{M}{\Delta t_d - d/c} = \frac{M}{\Delta\tau_d}\sqrt{\frac{1+\beta}{1-\beta}}\,, \tag{4.62}$$

which again yields Eq. (4.61).

These results show an asymmetry between the observations of the observer on Earth and the two astronauts O' and O".

- The observers O' and O" receive different numbers of signals during their respective observation time intervals, but the lengths of the time intervals are the same.
- Observer O receives the same number of signals from both observers O' and O", but during time intervals of different lengths.

At this point, a remark on the confusion that originated in the misinterpretation of the reciprocity in Eqs. (4.46) and (4.47) for the time dilation is in order. The misinterpretation resulted in the so called "twin paradox", which in its common form is stated as follows:

- A space traveler, who leaves Earth at some time, travels with relativistic velocity and finally arrives on the Earth again, will realize that he aged less than his twin who stayed on Earth.

It is sometimes argued that the twin on Earth could be the one doing the trip in opposite directions since motion is relative and the reference can be chosen arbitrarily. The confusion is related to faulty applications of the view of the local observer and the view of the frames of reference. To compare their ages, the individual twins must meet again at one place in space sometimes after their journeys. Thus, at least one twin must change its inertial frame of reference, and the equations for the time dilation do not apply in this simple form anymore.

The use of a more realistic scenario like that of a uniformly accelerated astronaut to α Centauri and back to Earth is discussed in Section 7.3.3. In Section 7.3.2 a scenario that compares the clock in a spacecraft on a circle with that of a non-accelerated observer on the periphery. If both twins are accelerated between two encounters the time intervals on their clocks can also be different.

5 Limits of the special theory of relativity

In fact, the special theory of relativity is a theory of empty space. If there are objects with mass and energy in space new hypotheses are required. In this Chapter we discuss the influence of masses and their mutual attraction or gravitation, which eventually leads to the theory of general relativity, which was predictive for phenomena like gravitational lensing, the perihelion precession of Mercury's orbit, gravitational waves or black holes. The two theories obey the correspondence principle, which means that they agree in the limit of small mass and energy densities.

5.1 Equivalence principle

We know from various experiments and experiences that, for example, an astronaut in a spaceship orbiting the Earth is weightless, or people in an airplane flying through an oblique parabola are weightless. This fact of weightlessness does not depend on the material, the mass and other structural qualities of the bodies in question. In a gravitational field, all bodies fall in the same way no matter what they are made of. A feather and a piece of lead accelerate the same in a vacuum tube, as is often demonstrated in school physics.

In classical mechanics, we have the second law of Newton,

$$F = m_{\mathrm{i}}\, a, \tag{5.1}$$

where F is the force acting on the body with inertial mass m_{i} that accelerates this body with acceleration a. Furthermore, we have the Newtonian law of gravitation,

$$F = G\,\frac{m_1\, m_2}{r^2}, \tag{5.2}$$

where F is the gravitational force that the two gravitational masses m_1 and m_2 mutually exert on each other if their distance is r, and $G = 6.6743 \cdot 10^{-11}\ \mathrm{m^3\ kg^{-1}\ s^{-2}}$ is the gravitational constant.

The relationship between the inertial masse and the gravitational mass is not given by classical mechanics or by special relativistic mechanics, but must be determined from other sources. Einstein stated in 1907 the equality of the inertial mass and the gravitational mass, and that the action of a gravitational field on a body is independent of the nature of the body. This is equal to the statement that the gravitational force as experienced locally by standing on a planet is not distinguishable from the force experienced in a linearly accelerating laboratory or rocket. This is called the weak or *Galilean equivalence principle*, which was proven to have an accuracy of 10^{-15} in the satellite experiment MICROSCOPE (Touboul, 2022). The weak equivalence principle was extended to the *Einstein equivalence principle* stating that the outcome of a local non-gravitational experiment in a freely falling laboratory is independent of its velocity and location in space.

https://doi.org/10.1515/9783111503592-006

The *strong equivalence principle* even allows for gravitational experiments: Specifically, an astronaut in a spacecraft who is exposed to gravitation does not feel the acceleration of resulting curved motion and could say that he is freely falling in the gravitational field. This would be true for all bodies in the spacecraft, thus it is not a force that is responsible for the motion. A motion without force is, according to Newton's first law, a uniform motion on a straight line. However, the astronaut knows that he is orbiting the Earth on an ellipse. As a curved ellipse and a straight line are not the same, this has to be reconciled. The contradiction can be resolved with the concept of *geodetic* lines. A geodetic line is the closest connection between two points in flat or curved space, or here the spacetime.

In this interpretation gravitation is not a force field. From a classical point of view, gravitation is an acceleration field since all massive bodies follow the same acceleration in the same point of space, independent of their mass, shape, chemical composition or any other observable qualities. In Section 5.2 we show that equal acceleration leads to equal curvature of spacetime.

5.2 Curvature of spacetime

If three points on a path do not lie on a line, the path is curved. In spacetime the simplest curves involve acceleration as they are e.g. encountered by a mass in a gravitational field. In the following we will treat free fall and the rotation of a planet in 1+1 and 1+2 spacetime and we will find a curvature that depends on the local acceleration and the speed of light.

5.2.1 Free fall in gravitational fields

The perhaps simplest, but not trivial situation is the free fall in a homogeneous field of gravitation. We assume an acceleration of a particle due to a gravitational acceleration g parallel to the z-axis of a Cartesian coordinate system, starting with velocity zero at the point $z = 0$. Thus the world line or spacetime trajectory

$$z = \frac{g}{2c^2}(ct)^2. \tag{5.3}$$

follows.

The world line of this body is a parabola with vertex at the origin $(z, ct) = (0, 0)$ (Figure 5.1). The osculating circle in the vertex of this parabola has a radius

$$r_g = \frac{c^2}{g}, \tag{5.4}$$

which again only depends on the speed of light, which is a universal constant, and g, which defines the strength of gravitation.

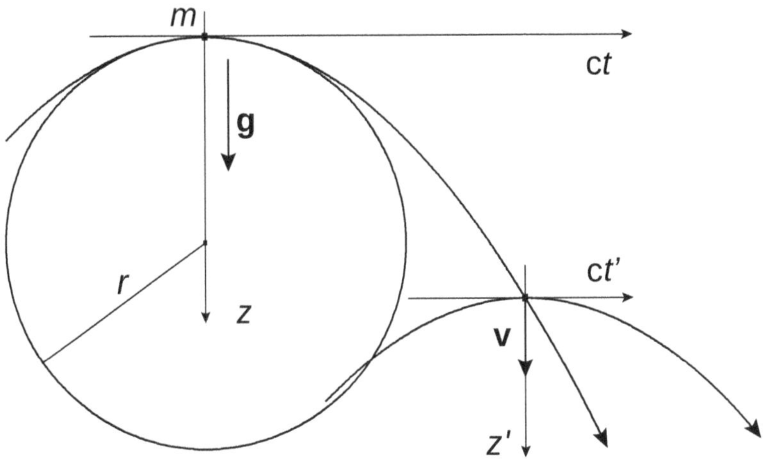

Fig. 5.1: World line of a mass *m* falling freely in a homogeneous field of gravitation with acceleration **g**. The world line in a coordinate system (*z*', *ct*') moving with the momentary velocity of the mass is congruent to the world line in the coordinate system (*z*, *ct*).

This universal constant is also found in the case at any other point of the world line with respect to a momentary co-moving inertial frame of reference. The transformation of Eq. (5.3) into the co-moving frame yields the same equation again. According to the principle of relativity, all inertial frames of reference are equivalent. Furthermore, this not only applies to homogeneous fields but also to spatially variable fields of gravitation. This implies that any situation of a freely falling body in a gravitational field locally follows a spacetime trajectory in the vicinity of the vertex of a parabola in the momentarily co-moving inertial frame of reference. This is true for any motion, even if it is highly relativistic with respect to some other inertial frame of reference. If this were not the case, some inertial frame of reference would be singular, and thus would violate the principle of relativity.

5.2.2 Planetary orbits

We consider a non-relativistic rotation with velocity v of a planet on an orbit with radius r around a star with mass M. Thus

$$g_r = \frac{v^2}{r} = G\frac{M}{r^2} \tag{5.5}$$

is the centripetal acceleration keeping the planet on a circular orbit in space.

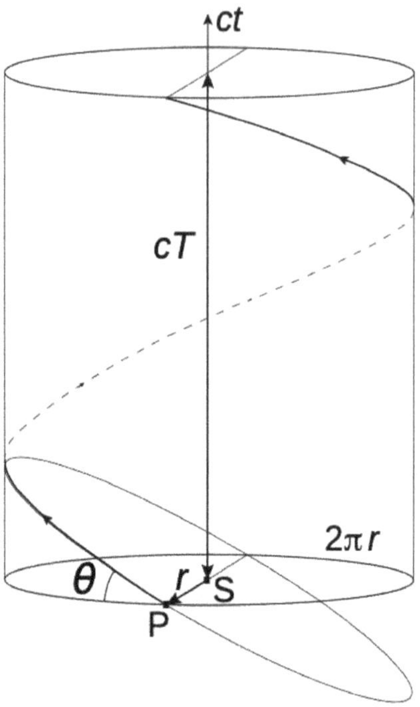

Fig. 5.2: Helical world line of a planet P orbiting a star S. The osculating ellipse at the position P (vertex of semiminor axis) has the same curvature as the world line in P. $c\,T$ is the distance along ct in one orbit period T.

The spacetime trajectory of this circular motion is a helical path around the mantle of a cylinder with spatial radius r (Figure. 5.2). In one revolution of duration T the helix climbs a distance $c\,T$ in the direction of the time axis, which is the symmetry axis of the cylinder. The pitch angle θ between the orbital plane in space and the spacetime direction of the momentary orbital motion is given by

$$\tan\theta = \frac{c\,T}{2\pi\,r}\,. \tag{5.6}$$

The momentary direction of motion and the spatial radius (planet-star) define a plane in 1+2 spacetime. This plane and the cylinder intersect in an ellipse and the momentary location of the planet is the vertex of the semiminor axis of this ellipse. The osculating circle of this vertex is also the osculating circle of the local spacetime trajectory, thus its radius also corresponds to the bending radius of the local world line.

The semiminor axis of the ellipse corresponds to the radius of the planetary orbit, $l_2 = r$. The semimajor axis l_1 then is

$$l_1 = \frac{r}{\cos\theta} = r\,\sqrt{1+\tan^2\theta}\,. \tag{5.7}$$

For planetary orbits, cT is of the order of light-years whereas the spatial circumference of the orbit is of the order of light-minutes to light-hours, thus $\tan\theta \gg 1$, and to a very good approximation

$$l_1 \approx r \tan\theta. \tag{5.8}$$

The radius r_g of the osculating circle in the vertex of the semiminor axis of the ellipse is

$$r_g = \frac{l_1^2}{r}. \tag{5.9}$$

Eliminating the angle θ from Eqs. (5.6) and (5.7) and eliminating the orbital period T with Keplers 3rd law,

$$T^2 = r^3 \frac{4\pi^2}{Gm}, \tag{5.10}$$

from Eq. (5.6) yields

$$r_g = \frac{c^2}{g_r}, \tag{5.11}$$

which is the bending radius of the planetary world line. This is the same bending radius as for the motion in a homogeneous gravitation field, Eq. (5.4).

In units of light-years (ly) and years (a), the speed of light is 1 ly a^{-1} and the acceleration near the surface of the Earth is $g = 1$ ly a^{-2}, which corresponds about to $g = 10$ m s^{-2} and a radius $r_g = 1$ ly, thus spacetime on Earth has a very small curvature, compared to her radius.

5.3 General relativity

In the two examples of movement in gravitation fields (Sections 5.2.1 and 5.2.2), the bending radii of the world lines only depend on the local acceleration g of gravitation. In the first example motion parallel to the direction of the gravitational field and in the second example a planetary orbit with a motion normal to the gravitational field are treated. In both cases, except for extreme stars with high densities such as neutron stars, the temporal component of the world lines make the bending radius very large.

Both motions can be considered to be free falls in the gravitational field, no additional external forces are necessary for the acceleration. If the moving bodies would be spacecrafts, astronauts would fall along the same path and thus would feel weightless. This corresponds to the equivalence principle, which states that the gravitational mass is equivalent to the inertial mass.

In the non-relativistic or special relativistic world, a free falling body would follow a straight line. In a gravitational field, the analog of a straight line is a geodetic line in spacetime, that is, the shortest world line connecting two points in spacetime. With these considerations, gravitation loses its status as a *force field* and re-emerges as a geometry of spacetime, and the curvature of spacetime takes the role of acceleration. A motion in a gravitational field becomes a *free fall*. In this picture gravitational mass does not exert a force on other masses such as the electric charge does on other electric charges. Gravitational mass acts on spacetime and spacetime acts on other masses.

The mathematical formulation of a theory of gravitation must connect the curvature of space to the spatial and temporal distribution of mass and energy. Curved space is described by Riemann geometry, where the metric is basic for the description of curvature. The generalization of the straight trajectory of force free motion in a space without gravitation is the geodesic in a spacetime with gravitation, which connects two points in spacetime along the shortest path in the given metric.

The mathematical description of Riemann spaces requires tensor calculus to describe metric and curvature, such as the *covariant* formulation of special relativity does for the energy-momentum tensor. Similar to electrodynamics, the mathematics restricts the number of possibilities. In the simplest formulation, two tensors **A** and **B** are related such that they are collinear,

$$\mathbf{A} = C\,\mathbf{B}, \tag{5.12}$$

where C is a scalar quantity. Since the 2^{nd}rank energy-momentum tensor is symmetric and divergence free, collinearity requires that the corresponding tensor that describes the metric of spacetime must be symmetric and divergence free, too. There is only one tensor with the above qualities that can be formed with the components and at most their second derivatives of the metric tensor, and this is the most likely candidate for the theory of gravitation. This tensor is now appropriately called the *Einstein tensor*.

Comprehensive presentations of the details of the theory of gravitation, or the *general theory of relativity*, can be found in Einsteins original work and many books (Einstein, 1916; Misner et al., 1973; Sexl and Urbantke, 1983; Schröder, 2007).

5.3.1 Einsteins tests of general relativity

The general theory of relativity was a radical change in paradigm concerning gravitation. Gravitation was not a force anymore but became a curvature of spacetime and bodies moving in curved spacetime followed so called geodesic lines. Of course, such a change in the theoretical foundation of gravitation requires empirical tests, and since the theory is so homogeneous and closed in itself, it needed to pass each singular test. It would be very difficult or impossible to correct such a theory in some details.

Einstein proposed three tests of general relativity (Einstein, 1916):

- The deflection of a beam of light when it passes close to a star
- The precession of the perihelion of an elliptic planetary orbit
- The red or blue shift of a beam of light in a gravitational field.

The prediction of the theory was that a beam of light grazing the sun is deflected by 1.75 arcseconds. It was during the solar eclipse on 29 May 1919 that a field of stars was photogaphed around the sun, and again later after the sun moved to another place. The small value of 1.75 arcseconds is difficult to determine but the results were in the right order of magnitude. At that time it was not completely clear whether they supported the old value of 0.875 arcseconds as predicted by a theory using Newtons gravitation law or general relativity.

It was already clear at that time that the perihelion of planet Mercury is precessing due to the perturbation of the other planets. Its value was known to be 531 arcseconds per century. However, the observed precession is 43 arcseconds per century too large to be explained by perturbation due to the other planets. This difference could be exactly explained by the general theory of relativity. It is remarkable to what accuracy the observed value was known, to what accuracy celestial mechanics was able to compute the value without computers, such that the 43 arcseconds per century could be confirmed to be a general relativistic correction.

Long before Einstein published his general theory of relativity (Einstein, 1916), he predicted the influence of gravitation on the frequency of light (Einstein, 1907, 1911). He wrote:

> ... daß von der Sonnenoberfläche kommendes Licht, welches von einem solchen Erzeuger herrührt, eine um etwa zwei Millionstel grössere Wellenlänge besitzt, als das von gleichen Stoffen auf der Erde erzeugte Licht (Einstein, 1907).
> ... that light from the sun's surface, which comes from such a generator, has a wavelength that is approximately two millionths longer than the light produced by the same substances on Earth.

The white dwarf Sirius radiates at a temperature of 8000°C and has absolute magnitude of 11.3, which gives a radius of 19'600 km and a density of 53'000 kg m^{-3}, if a mass of 0.85 solar masses are assumed (Eddington, 1924). To clarify these uncertainties, Eddington proposes in his paper to measure the Einsteinian (red) shift of the spectrum. Adams (1925) accurately measured the spectrum of Sirius and confirmed Eddingtons prediction of the density of Sirius. Since the gravitational redshift is very small within our solar system or in an Earth bound laboratory, it was only experimentally verified about 35 years later by Pound and Rebka Jr. (1959). The experiment was performed in the tower in the Jefferson laboratory at Harvard University. They measured the predicted frequency shift with the Mößbauer effect where they used radioactive ^{57}Co, which emits γ-radiation, in a 22.5 meter high tower.

5.3.2 Black holes: horizons and singularities

Black holes are massive objects with horizons and predicted singularities. If a light source is invisible for an observer, it is behind a horizon. If the curvature of spacetime diverges we deal with a singularity.

The calculation of curvature of planetary orbits may not be true anymore for very massive stars and very close orbits with short orbital periods. We expect relativistic corrections as soon as the planetary speed approaches the speed of light. From Eqs. (5.5) and (5.6) it is seen that this is the case if the radius of the planetary orbit r approaches the space curvature r_g, i.e. for $\sqrt{r/r_g} \to 1$.

A white dwarf star with a mass $m = 2.8 \cdot 10^{30}$ kg (1.4 times solar masses) and a density $\rho = 10^9$ kg m^{-3} has a radius $r = 8.7 \cdot 10^6$ m, and on its surface a gravitational acceleration $g = 2.5 \cdot 10^7$ ms^{-2}. Thus the bending radius of a trajectory of a particle orbiting the white dwarf near its surface is $r_g = 3.6 \cdot 10^9$ m.

An even more extreme example is a neutron star with a mass $m = 4 \cdot 10^{30}$ kg (2 solar masses) and a mean density $\rho = 5 \cdot 10^{17}$ kg m^{-3}. Its radius is about $r = 1.3 \cdot 10^4$ m, its acceleration near its surface $g = 2 \cdot 10^{12}$ ms^{-2} and the bending radius near its surface is $r_g = 4.5 \cdot 10^4$ m.

To calculate the orbital velocities near the surfaces of white dwarfs and neutron stars, we combine the classical Newtonian law of gravitation with the relativistic acceleration of circular orbits, Eq. (3.32),

$$\gamma^2 \frac{v^2}{r} = \frac{G\,m}{r^2}, \tag{5.13}$$

which gives

$$\beta^2 = \frac{v^2}{c^2} = \frac{G\,m}{r\,c^2 + G\,m}. \tag{5.14}$$

This results in $\beta = 0.014$ for white dwarfs and $\beta = 0.82$ for neutron stars, the latter corresponds to $\gamma = 1.7$. This means that the above non-relativistic determination of the bending radius is still accurate for white dwarfs (100 ppm) but not anymore for neutron stars (75%).

It is not clear whether the equation for gravitation still holds for such massive and small stars. The physical processes that counter the gravitational force are limited. Once nuclear fusion in a star stops, the radiation pressure fades and it starts to shrink until the pressure of the remaining matter balances gravity. Stars with a mass of up to 1.4 solar masses become white dwarfs with a diameter of about Earth-size. Stars with masses of about 10 to 25 solar masses shrink to a few kilometers after loosing the largest part of their mass. Now, the so called neutron degeneracy counteracts gravitation with the Pauli principle that does not allow for more than two spin 1/2 Fermions, here neutrons, at the same "point". Neutron stars with more than 2 solar masses are not stable, loose the Fermionic repulsion and further implode. The end product is a so-called *black hole*. The general theory of relativity gives the Schwarzschild radius around a collapsed non-rotating star in Eq. (5.15), which is an *apparent horizon*, through which no light can escape. It is sometimes speculated that the collapsed stellar mass shrinks to a singularity at the center of a black hole. However, there is no experiment to test this hypothesis.

5.3.3 Classical approaches to general relativity

Three astronomical topics were treated long before Einstein published the general theory of Relativity. (i), the possible existence of dark stars because of the limited escape velocity of light by Michell (1784) and later independently by Laplace (1796), (ii) the deflection of light due to gravitational attraction when it passes close to a star by Cavendish (1784) and von Soldner (1804), and even Einstein (1911), and (iii) the precession of the perihelion of a planetary orbit due to a finite speed of the propagation of a change of gravitation by Gerber (1898). Interestingly, all three cases, although treated purely within classical physics, yield results very close to the results of general relativity. We only present their results, but not the derivations of the results.

(i) Michell (1784) and Laplace (1796) assumed a corpuscular (particle) picture of light. They computed the radius of a star with mass m such that the escape velocity on its surface is the speed of light. They considered the speed of light c as variable and assumed that it would decrease as it moves out of the gravitational field. This radius r of such a star with mass m and a gravitational constant G is

$$r = \frac{2\mathrm{G}\,m}{c^2}, \tag{5.15}$$

which corresponds exactly to the Schwarzschild radius (Schwarzschild, 1916) of a black hole. Although the two radii are the same, it has to be noted that they were obtained with two different theories and two different pictures of light. It was speculated whether this is a pure coincidence or if there is some deeper meaning in it (Preti, 2009). We can not answer this question, but we can outline an argument for this *coincidence* by applying a dimensional analysis.

Tab. 5.1: Dimensional matrix: D is the dimensional quantity, L is length, T is time and M is mass. The physical quantities are the gravitational constant G, a mass m, a velocity c and a distance r. The numbers in the table are the exponents of the dimensional quantities of the corresponding physical quantities.

D	G	m	c	r
L	3	0	1	1
T	-2	0	-1	0
M	-1	1	0	0

An equation is called dimensionally homogeneous if it does not depend on the choice of the fundamental units. As a consequence such equations can be written in the form

$$1 = f(d_1, \cdots, d_n) \tag{5.16}$$

where f is a function of the dimensionless combinations d_i of physical variables in this equation, see e.g. Hutter and Jöhnk (2004). We now can define a dimensional matrix, where the lines contain the exponents of the corresponding dimensional quantity and the columns represent the corresponding physical variables. For the given example, the relevant physical variables are the gravitational constant G, the mass of the star m, the escape velocity c and the radius in question r. The dimensional quantities are the length L, the time T and the mass M, see Table 5.1.

The Buckingham theorem (Buckingham, 1915) states that the number of dimensionless combinations of the given physical quantities is equal to the number of physical quantities minus the rank of the dimensional matrix. In our case, we have 4 quantities and the rank of the matrix is 3, thus we have exactly one dimensionless combination,

$$1 = \frac{\mathrm{G}\, m}{c^2\, r}, \tag{5.17}$$

and any solution of the problem must contain this combination. For the Schwarzschild radius, Eq. (5.15) or the radius of a dark star according to Michell (1784), we split the dimensionless quantity to

$$r = \frac{\mathrm{G}\, m}{c^2}, \tag{5.18}$$

which is the simplest possibility. The Schwarzschild radius is obtained up to a constant factor, which can not be predicted by dimensional analysis. The choice of the variables in question for the Schwarzschild radius and for the radius of a star, where the escape velocity corresponds to the speed of light c, are the same, even with the same constant factor of 2.

(ii) Perhaps inspired by Michell (1784) to treat light as particles with a possibly variable velocity, Cavendish (1784) and later von Soldner (1804) treated the deflection of light passing close to a star (Will, 1988). The two results are not exactly the same, perhaps because Cavendish (1784) assumed the speed of light to be c at infinity, whereas von Soldner (1804) started with c at the surface of a star (Will, 1988). Their values of the deflection of light when passing the sun at its surface is 0.875 arcseconds, which is half the value predicted by general relativity. Also Einstein (1911) made the same calculation and obtained the same value, however, he abandoned this idea as being not sufficient.

Cavendish (1784) only gives his result in a very short note whereas von Soldner (1804) presents a detailed derivation. He assumed the speed of light to be constant in the direction of its motion but integrated the gravitational acceleration normal to the motion, assuming that light behaves like a material body. The notation of von Soldner (1804) for the defection angle can be rewritten as

$$\Delta\vartheta = \frac{2\, G\, m}{c^2\, r}. \tag{5.19}$$

Considering the fact that about the same physical quantities as for the dark stars of Michell (1784) are involved in this deflection problem, this result is again not surprising. It corresponds to the above dimensional analysis of Eq. (5.17), where the additional quantity of the deflection angle $\Delta\vartheta$ is dimensionless, indicating again the limitation of the dimensional analysis if it is about obtaining the exact result.

(*iii*) The situation is similar in the case of the precession of the perihelion of the orbit of a planet around a star. Gerber (1898) computed such an orbit assuming that any change in the Newtonian gravitational field propagates with the speed of light. His result for the precession of the perihelion, ϵ, is equivalent to the one found by Einstein (1916) with the general theory of relativity,

$$\epsilon = \frac{6\pi}{1-e^2}\frac{G\,m}{c^2\,r}, \tag{5.20}$$

where e is the eccentricity and r is the semi-major axis of the orbit, G is the gravitational constant and m the mass of the star. In this equation the same physical quantities, G, m, c and r are used together with two dimensionless quantities, ϵ and e. Again, the same dimensionless combination as in Eq. (5.17) occurs in the equation. The way how ϵ and e and the factors 6 and π enter the equation can not be determined by dimensional analysis alone. It is surprising that a pure classical treatment except for the speed of propagation of gravitation yields the same result as general relativity. Whether this is coincidental or has a deeper meaning is a difficult question that we can not answer.

6 Experimental evidences

In this Chapter, we present a series of examples in daily life, technology and science, in which special relativity plays an important role. The presentation of examples that may be of interest to many people, either they occur in daily life such as the Global Positioning System (GPS), or in medical applications, such as x-ray imaging. Science fiction aspects and space travel are treated in detail in Chapter 7.

The four Sections treat time dilation, electromagnetism, electromagnetic radiation fields, and relativistic effects in atoms. Electromagnetism treats the forces between charges that are at rest or moving with respect to each other. It is, in principle more complicated than electromagnetic radiation. Once the radiation is created, and if it does not interact with matter, it exists in vacuum and does per se not rely on mass and charge. The relativistic effects in atoms due to the motion of the electrons can be measured precisely from electromagnetic radiation emitted or absorbed by the atoms.

6.1 Time dilation

Time dilation is one of the most intriguing consequences of special relativity. Here we recall the fact that cosmic muons reach the sea level thanks to time dilation, we discuss the implications of time dilation for the Global Positioning System (GPS) and finally summarize the results of the Haefele-Keating experiment. All three examples display isotropic time dilation like in Eq. (4.55).

6.1.1 Cosmic radiation: muons

Muons are elementary particles that are created by cosmic rays hitting the Earth atmosphere. They have a characteristic rest mass, charge, and magnetic moment, and they have a finite lifetime. The lifetime in the rest frame of the particles is, due to time dilation, not the same in a moving frame. This has the consequence that relativistic muons which are created in the stratosphere may reach sea level.

Cosmic rays mainly consist of highly energetic protons with energies of up to 10^{20} eV. At a height of about 20 km, the earth's atmosphere is dense enough, such that the protons interact under the formation of isotopes like carbon 14 and a shower of short-lived π mesons with lifetimes of less than 25 ns. The charged π mesons decay into relatively long-lived muons (μ^+ and μ^-). The muons have a much smaller interaction cross-section with the atmosphere than the primary cosmic rays, but still loose energy and decay after a lifetime of 2.2 μs into electrons or positrons and neutrinos. About one muon per square centimeter and minute arrives at sea level, while the flux is about twice as large at an altitude of 2 km.

https://doi.org/10.1515/9783111503592-007

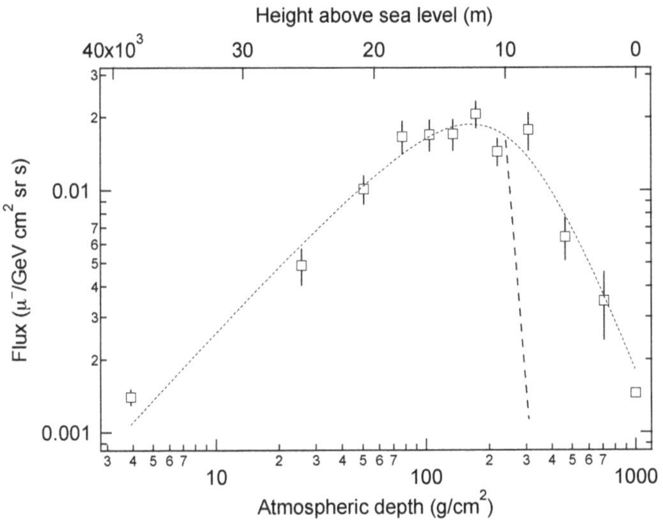

Fig. 6.1: Measured flux of muons in the earth's atmosphere as a function of the atmospheric depth, which relates via the barometric formula to the height. Data for μ^- with momenta between 0.3 and 0.53 GeV/c from Boezio et al (2000). The dotted line is a guide to the eye. The dashed line is the expectation for muons created at a height of 10 km and moving with the speed of light, without time dilation.

Figure 6.1 shows measured muon fluxes from the CAPRICE94 experiment where a muon detector was lifted in a balloon to a maximum height of 38 km above sea level (Boezio and 33 others, 2000). The data confirm that muon production starts in the topmost atmosphere, peaks about 15 km above sea level and then reduces by about a factor 15 down to sea level.

Given the lifetime $\tau_0(\mu)$ of 2.2 μs the fact that such a large number of muons arrives at the sea level can only be understood in the framework of the special theory of relativity. It is the example of time dilation, i.e. the need for the distinction in which frame of reference time and distance are measured. Here the distance of 15 km is measured in the reference system of the Earth and the lifetime is referenced to the rest frame of the particle.

For a muon with a momentum of 0.4 GeV/c we get with Eq. (3.43) and mass $m(\mu^-)$=106 MeV/c^2 a relativistic factor γ of about 4, and a corresponding reach or decay length in the atmosphere of $c\beta\gamma\tau_0$ of 2.5 km. Comparison with the observed attenuation length of 3.4 km indicates that dilated muon decay is the main reason for the decrease of the muon flux in the lower atmosphere as shown in Figure 6.1. The discrepancy remaining is assigned to the energy losses of the muons in the atmosphere and the related cascade processes, as can be observed in bubble chambers.

The explanation for the long decay time in the frame of reference of the Earth can be understood as well with the proper velocity ω of the muon (see Section 3.1), which is about 3.8 c for the discussed case.

6.1.2 Global navigation satellite system (GPS)

In recent years the Global Positioning System (GPS) or more precisely the Global Navigation Satellite Systems (GNSS) became an important tool for orientation on Earth. A GNSS receiver provides the information on a location in Minkowski space, i.e. three spacial coordinates (x, y, z) and the time coordinate (ct). Essentially it is a precise triangulation: from the distance to four transmitters, of which the positions are precisely known, the coordinates (x, y, z) may be determined with high accuracy.

Four satellite solution

Navigation systems comprise a number of transmitters, of which the positions are accurately known. The receiver at an unknown position P obtains the information on the distances to some of the transmitters by measuring e.g. the time of signal transmission. In a non-relativistic world, accurately synchronized clocks suffice to obtain a reasonable accuracy of the position of P.

The satellite-borne GNSS presently can be used to measure positions to an accuracy well below meters, with some additional correction even down to a few millimeters. Since the speed of light is so large, the required accuracy of the time measurements becomes quite demanding. An error of 10^{-9} seconds results in a position error of 0.3 meters. Present-day clocks achieve better accuracy over substantial time spans in the laboratory.

An observer P on Earth wants to obtain his accurate spatial position $\mathbf{r}_{s,P}$. The GNSS receiver at his position must measure the accurate time $t_{P,i}$ of the arrival of a signal from satellite i, which was transmitted from the satellite at time t_i at position $\mathbf{r}_{s,i}$,

$$|\mathbf{r}_{s,P} - \mathbf{r}_{s,i}| = c\,(t_{P,i} - t_i) \tag{6.1}$$

where the index $_s$ denotes spatial vectors. In relativistic terminology, the observer at position P collects the information from the retarded light cone, i.e. a time-like vector in spacetime (see Section 2.3.1),

$$|\mathbf{r} - \mathbf{r}_i| = 0 \tag{6.2}$$

where $\mathbf{r}$ and $\mathbf{r}_i$ are 4-vectors. In principle, three satellites are necessary for position measurement, if the positions of the satellites and the synchronized time at the moment of the transmission of the signal is known, together with the synchronized time at the position of the GNSS receiver. In practice, a GNSS device should be cheap and user-friendly, and thus it will be neither part of the global synchronization network nor will it contain an expensive clock. Thus the entire spacetime position vector

$$\mathbf{OP} = \mathbf{r}_{P,i} = \begin{pmatrix} t_{P,i} \\ x_P \\ y_P \\ z_P \end{pmatrix} \tag{6.3}$$

is unknown or only known to some rough approximation. This requires four satellites to determine the four unknowns of $\mathbf{r}_{\mathrm{P},i}$, the three spatial components and the synchronization Δt_P of the clock of the receiver,

$$t_{\mathrm{P},i} = t_{\mathrm{mP},i} - \Delta t_\mathrm{P}\,, \tag{6.4}$$

where $t_{\mathrm{mP},i}$ are the times of arrival measured with the clocks of the receiver.

With these measurements, four non-linear equations for four unknowns can be written,

$$|\mathbf{r}_{\mathrm{s,P}} \cdot \mathbf{r}_{\mathrm{s},i}| = c\,\left(t_{\mathrm{mP},i} - \Delta t_\mathrm{P} - t_i\right)\,. \tag{6.5}$$

The used solution procedure for the four equations applies a Taylor expansion about an approximate estimated position and the solution of the resulting linear system.

Special Relativity corrections

Maintaining a number of space-borne clocks in a synchronized state requires routinely performed corrections based on detailed monitoring of orbital elements. This not only concerns special relativistic but also general relativistic effects of gravitation onto clock rates. General relativity is not treated in more detail here, see e.g. Ashby (2002, 2003).

The accuracy of the GNSS is limited by the synchronization of the clocks. The special relativistic effect affecting the rate of moving clocks is time dilation. The clocks in different satellite orbits move in different momentary frames of reference. Satellites at a distance of about 20'000 km from the Earth surface move with a velocity of about 4000 ms^{-1} with respect to the center of the Earth. Due to time dilation, Eq. (4.45), the rates of two clocks deviate by

$$\frac{\Delta\nu}{\nu} = \frac{1}{\gamma} - 1 = \sqrt{1-\beta^2} - 1 \approx \frac{1}{2}\beta^2 \approx 10^{-10}\,, \tag{6.6}$$

thus every 30 seconds, the position error increases by about 1 meter. This effect is thus far too large to be ignored for the GNSS.

General Relativity corrections

The effects of smaller gravitation at the height of the orbits is a factor 4 to 6 larger than the special relativistic time dilation and is of the opposite sign. At the height of the GNSS satellites, the net general and special relativistic effects make the clocks fast with respect to clocks on Earth. This discrepancy is mostly compensated for by slowing the clock rate before the launch of the satellites.

To maintain the long-term accuracy of the GNSS, active maintenance of the synchronization is required. Many other factors must be considered, e.g. the changes in the orbital elements must be monitored and delays in the signals during the passage of the ionosphere can be corrected by using two different frequencies and exploiting the dependence of the delay on the frequency.

The chosen frame of reference is Earth-centered and rotates with a constant rotation rate. The task is then to synchronize the satellite clocks with respect to a rotating frame. The synchronization is performed with standard Einstein synchronization using electromagnetic waves. In a rotating frame, the so-called Sagnac effect must be considered, i.e. the signal time is advanced or retarded because the endpoints of the path move due to the rotation of Earth during transition time.

6.1.3 The Hafele-Keating experiment

In 1971, jet planes and atomic clocks became available, and this opened the possibility to test relativistic time dilation with macroscopic clocks for achievable velocities. Hafele and Keating (1972a) performed that experiment with portable cesium beam clocks on commercially available flights of jet planes, flying around the world in both, westward and eastward, directions. The clocks were compared to equal clocks at the U.S. Naval Observatory, MEAN(USNO). The resulting time shifts are given in the abstract of Hafele and Keating (1972a):

> Four cesium beam clocks were flown around the world on commercial jet flights during October 1971, once eastward and once westward, recorded directionally dependent time differences which are in good agreement with predictions of conventional relativity theory. Relative to the atomic time scale of the U.S. Naval Observatory, the flying clocks lost 59±10 nanoseconds during the eastward trip and gained 273±7 nanoseconds during the westward trip, where the errors are the corresponding standard deviations. These results provide an unambiguous empirical resolution of the famous clock "paradox" with macroscopic clocks. Hafele and Keating (1972a)

The predicted values of the corresponding time shifts were 40±23 nanoseconds during the eastward trip and 275±21 nanoseconds during the westward trip (Hafele and Keating, 1972b). Unfortunately, Hafele and Keating (1972a) mention the *famous clock "paradox"* in the abstract, which sometimes leads to the faulty assumption that they confirmed the time dilation of special relativity. Hafele (1970) and Hafele (1972) derived the theory on the basis of the general theory of relativity. The applied approximations, weak gravitational field and slow velocities, lead to equations that resemble special relativistic equations. Higher-order terms would not have been observable with the accuracy of the cesium beam clocks at that time. Furthermore, Hafele (1972) states

> Notice that this question does not ask about Doppler shifts or other instantaneous effects between the two clocks. No signals are transmitted between the clocks during the flight.

The clocks were only compared at the same place and at the same times before and after the flights. It is only assumed that the clocks record proper time.

6.2 Electromagnetism

Magnetic fields and their action on moving charges are probably the most far-reaching consequence of special relativity. The Maxwell equations show that magnetic fields are a direct consequence of the finite speed of light (Section 3.3).

According to the third Maxwell equation, Eq. (3.47), a temporal change in the magnetic field produces a curl in the electric field, which, under certain circumstances may produce an electric current. This law is called Faraday's law of induction. Reversely, the fourth Maxwell equation, Eq. (3.48), states that a temporal change in the electric field or an electric current produces a curl in the magnetic field. Both effects are widely applied in a series of technical devices and household appliances.

6.2.1 Voltage and current

Electric and electronic devices are an application of relativity. To illustrate this with appealing examples, first a primer in electricity with the principles and terminology is given.

Eq. (3.49) defines the electric current density through the charge density and the velocity of the charges as a vector field in space and has the dimension of *charge per time per area.* Usually, in electric devices, the current is confined to some electrically conducting material of a certain geometry, mostly wires with a given cross-sectional area, and the *electric current* I is defined as the total charge Δq moving with an average velocity v through a cross-section A in a given time Δt, see Figure 6.2,

$$I = \frac{\Delta q}{\Delta t} = v\,\frac{\Delta q}{\Delta s}\,. \tag{6.7}$$

In metric units, electric current is measured in *Ampère*, A=C/s (Coulomb per second), where 1 C corresponds to the electric charge of 6'241'509'629'152'650'000 electrons, thus each electron carries a charge of $-1.602 \cdot 10^{-19}$ C.

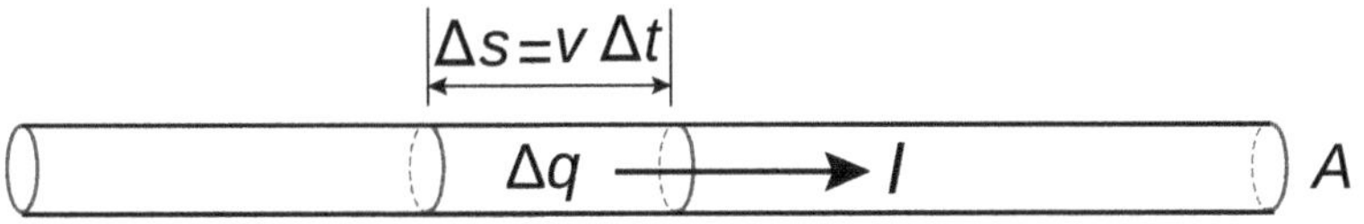

Fig. 6.2: Schematic of a current I in a wire with cross-section A. The charge Δq moves with velocity v (see Eq. (6.7)).

To generate an electric current, charged particles (mostly electrons) must be accelerated by an electric field through the Coulomb force, Eq. (3.81). The mathematical formulation of the force on a charged particle with charge q in an electric field of strength $\mathbf{E}$ is equivalent to the classical formulation of the force on a mass m in a gravity field $\mathbf{g}$,

$$\mathbf{F}_C = q\,\mathbf{E}\,, \quad \mathbf{F}_g = m\,\mathbf{g}\,, \tag{6.8}$$

where $\mathbf{F}_C$ and $\mathbf{F}_g$ are the Coulomb force and the weight, respectively. To define the potential of an electric field, constant electric and gravity fields along the lifting direction are compared. To lift a mass m to a height Δh in the gravity field $\mathbf{g}$ requires an energy $E_{\text{pot,g}} = mg\,\Delta h$. Correspondingly, to lift a charge q in an electric field $\mathbf{E}$ an energy $E_{\text{pot,C}} = qE\,\Delta h$ is required. These energies are called *potential energies*, and the potential energy per charge or per mass is called *potential* with respect to Δh=0,

$$U_C = \frac{E_{\text{pot,C}}}{q}\,, \quad U_g = \frac{E_{\text{pot,g}}}{m}\,. \tag{6.9}$$

In the case of static fields where the field strength and direction may be variable, the potential difference between two points is calculated with the path integral of the force vector, which is not dependent on the chosen path. Such fields are called *conservative fields*. Importantly, the potential is always defined as a potential difference, or the potential at one point is only determined up to an additive constant. In the metric system of units, an electric potential difference is measured in Volt, V=J/C (Joule per Coulomb).

In an electric field in a vacuum, an electric charge q is accelerated along the field vectors. If started in a given reference system at zero velocity, its kinetic energy at an endpoint in this system corresponds to the potential energy difference

$$E_{\text{kin}} = q\,U_C\,. \tag{6.10}$$

This kinetic energy defines the energy unit *electronvolt*, eV, which a particle with the electric charge of one electron attains if it runs through a potential difference of one Volt,

$$1\ \text{eV} = 1.602 \cdot 10^{-19}\,\text{J}\,.$$

In an electrically conducting material, the electric field accelerates the charge, up to a scattering event, where the charge loses the kinetic energy to the heat bath of the conductor. In equilibrium the charges move with an average drift velocity that is proportional to the current density, where the charge carrier density is the proportionality constant (Eq. (3.49)).

The work done by the electric field is dissipated as the so-called Joule heat E_J,

$$E_J = q\,U_C\,. \tag{6.11}$$

If not one particle but an electric current I is considered, the corresponding Joule heat rate (Ohmic heating) or power dissipation is

$$P_J = I\,U_C\,. \tag{6.12}$$

Finally, we find Ohm's law: for a given material and geometry, the electric current I is proportional to the potential difference U_C applied,

$$I = \frac{U_C}{R}, \tag{6.13}$$

where R is the electric resistance of the device, measured in Ohm, Ω=A/V (Ampère per Volt).

6.2.2 Generators and Motors

Electric motors and generators are commonly referred to as electric machines. An electric motor converts electrical energy into mechanical energy, while an electric generator does the reverse: conversion of mechanical energy to electrical energy. An electric current I in a wire produces a magnetic field, but also interacts with a magnetic field. Charge moving with velocity $\mathbf{v}$ relative to a magnetic field is deflected by the Lorentz force $\mathbf{F}_L$, Eq. (3.82). If it is confined in a wire, it exerts a force on the wire as a whole,

$$\mathbf{F}_{\text{wire}} = q\,(\boldsymbol{v} \times \mathbf{B})\,. \tag{6.14}$$

where $\mathbf{F}_{\text{wire}}$ is the force acting on a piece of wire of length Δs due to a magnetic field of strength B normal to the axis of the wire, and q is the moving charge within this part of the wire (Figure 6.2).

In many applications, the magnetic field vectors point normal to the wire axis, thus

$$F_{\text{wire}} = qv\,B = I\,B\,\Delta s\,. \tag{6.15}$$

This is the force that drives electric motors. Each moving charge produces a magnetic field that exerts a Lorentz force on other differently moving charges. Thus, a wire carrying a current I_1 feels a force in the magnetic field of a second wire carrying a current I_2. We only consider the case of parallel wires, thus the wire with current I_1 feels a force, Eq. (6.15), with a magnetic field produced by the current I_2, Eqs. (3.74) and (3.78), thus

$$F_{\text{wire}} = \frac{I_1 I_2}{2\pi\epsilon_0 d}\,\Delta s\,, \tag{6.16}$$

where d is the distance between the wires. The force is attractive if the currents in the two wires flow in the same direction, otherwise, it is repulsive. This force can be demonstrated with two parallel wires carrying an electric current, and is perhaps one of the earliest experiments with electromagnetic forces performed by André-Marie Ampère in 1820.

Eqs. (6.15) is the base of an electric motor, where a current in a magnetic field evokes a force or an electric generator, where a force on a conductor evokes a current. For a cyclic electric machine we distinguish a rotor and a stator. The rotor rotates on an axis and an electrical current generates the torque and vice versa. Most electric motors operate through the interaction of magnetic fields and current-carrying conductors to generate mechanical force. The reverse process, producing electrical energy from mechanical energy, is done by generators such as an alternator or a dynamo. Some electric motors can also be used as generators, for example, a traction motor on a vehicle may perform both tasks.

Electric motors have one thing in common: the force that drives to rotate one part of the motor is the Lorentz force that the moving electrons in the electric current feel if they are exposed to an external magnetic field. The various types of motors are different in the way the magnetic field is produced, by permanent magnets or by electric coils. Another difference is the way in which the Lorentz force works to rotate the motor, either on currents in wires and coils or on currents in rotating discs. In conventional motors with wires and coils, the polarity of the magnetic field must be adjusted adequately during one revolution. In so-called unipolar motors, no adjustment of the magnetic field is required. Conceptually this is considered to be simpler and we thus treat the following unipolar electric machines that base on the Barlow wheel.

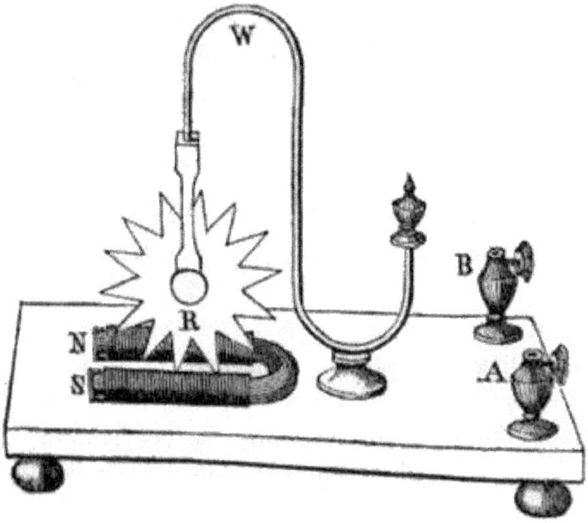

Fig. 6.3: Scheme of Barlow's unipolar motor as depicted in the Manual of Magnetism (Davis et al., 1842). The electric current flows on a radius (R) of the star-shaped wheel between the axis (connected to pole A) and liquid mercury in the groove of the stand (connected to pole B). The source of the magnetic field is indicated by the north- and south poles (N and S) of the horseshoe magnet and points perpendicular to the current. The resulting Lorentz force produces a torque that causes the wheel to rotate with the direction of rotation depending on the direction of the current. Applying an external torque on the wheel induces a voltage between A and B and realizes a unipolar dynamo or DC generator.

In 1822 Peter Barlow built a machine that realized a unipolar electrical engine (Figure 6.3). This engine, known as Barlow wheel, uses the Lorentz force for locomotion.

The engine is called unipolar since the current direction does not need to be changed during the continuous rotation. A current is flowing along the radius from the wheel axis to the liquid mercury contact at the wheel periphery. Together with the magnetic field, this imposes a Lorentz force on the charge carriers, which induces a torque and lets the wheel to rotate. In 1831 Faraday used such a device in order to demonstrate a DC current source by rotating the wheel and measuring a current between pole A and pole B.

Sometimes, the Barlow wheel is called a *true relativistic* (Guala-Valverde et al., 2002) motor. Since any electromagnetic device implies the special theory of relativity, via the Lorentz force, the Barlow wheel is just an example of a specific design of such a motor. Any electromagnetic motor is relativistic.

6.3 Electromagnetic radiation fields

Electromagnetic radiation is omnipresent in our world and in particular in daily life. In astronomy most information stems from the observation of electromagnetic radiation. It consists of waves or massless particles that carry energy and interact with matter. Our eyes are sensory organs that can sense electromagnetic waves and extract information on direction and frequency to interpret them as colored pictures in narrow frequency window. Electromagnetic radiation is used to transport information also in the invisible part of the electromagnetic spectrum, such as that of radio waves.

6.3.1 Definitions

Electromagnetic radiation transports energy from an emitter or source to an absorber or receiver. A radiation measuring device collects (absorbs) energy from a radiation source on a given sensor area. Electromagnetic waves obey the dispersion relation $c = \lambda\nu$, Eq. (3.59), where λ and ν are the wavelength and frequency, respectively, and c is the speed of light. Here we only consider electromagnetic waves in vacuum, thus the speed c corresponds to the phase velocity and is universally constant.

In the dualistic picture (Section 1.4.4), electromagnetic radiation can be considered as waves according to Eqs. (3.57) and (3.58), but also as particles (photons) carrying energy $E = h\nu$, Eq. (1.5).

Radiation may be received from a radiation field from parts or the entire celestial (directional) sphere, or from discrete sources. To describe such a radiation field or source, we define several quantities that play a role for different purposes (Table 6.1).

The *radiant energy* H is the total energy emitted by a given source or received by an absorber. The *radiant flux* or *intensity* J is the radiant energy per unit time and unit area,

$$J \equiv \frac{\mathrm{d}H}{\mathrm{d}A\,\mathrm{d}t}\,, \tag{6.17}$$

measured in units [$\mathrm{Jm^{-2}s^{-1}}$] or [$\mathrm{Wm^{-2}}$].

Tab. 6.1: Glossary of different quantities related to impact of radiation.

quantity	symbol	unit	definition
radiant energy	H	J	energy
radiant power	P	W	power
radiant power intensity	I	$\mathrm{W\,sr^{-1}}$	radiant power per unit solid angle
radiant flux, intensity	J	$\mathrm{W\,m^{-2}}$	radiant energy per unit time and area
irradiance	J	$\mathrm{Wm^{-2}}$	power incident on a surface
radiant emittance	J	$\mathrm{W/m^{-2}}$	power emitted from a surface
radiance	L	$\mathrm{Wsr^{-1}m^{-2}}$	power per unit solid angle per unit projected source area

A plane surface receives radiation from all directions of the "celestial sphere" above the plane. The apparent area or field of view of a radiating element of the celestial sphere is called *solid angle*, which is 2π for the half sphere visible above a given surface. The *radiance* L, depends on the direction, defined by the polar angle θ between the direction to the source of the radiation and the normal vector to the surface (Figure 6.4),

$$L \equiv \frac{\text{potential irradiance}}{\text{solid angle}} = \frac{\Delta J_\theta}{\Delta\Omega} = \frac{\Delta J}{\Delta\Omega\cos\theta}\,, \tag{6.18}$$

measured in units [$\mathrm{Wm^{-2}sr^{-1}}$] (sr=steradian). ΔJ is the energy per area arriving on the surface in question (irradiance) and ΔJ_θ is the potential irradiance, if the surface were oriented (with its normal vector) towards the solid angle element from where the radiation is coming.

$$L \equiv \frac{\mathrm{d}J}{\mathrm{d}\Omega\cos\theta} = \frac{\mathrm{d}H}{\mathrm{d}A\,\mathrm{d}t\,\mathrm{d}\Omega\,\cos\theta}\,. \tag{6.19}$$

The same quantities may be defined for given spectral ranges.

An often-considered situation in a radiation measurement concerns a horizontal plane, e.g. a radiation sensor, with a given area that receives radiation over a given spectral range from the entire celestial sphere above the plane. The sensor receives the energy from a radiation field with an irradiance distribution of $L = L(\varphi, \theta)$, which depends on the azimuth φ and the zenith angle θ (Figure 6.4). The *differential irradiance* $\mathrm{d}J$ on the sensor from a solid angle element $\mathrm{d}\Omega = \sin\theta\,\mathrm{d}\theta\,\mathrm{d}\varphi$ is equal to

$$\mathrm{d}J = L\cos\theta\,\mathrm{d}\Omega = L(\varphi,\theta)\sin\theta\cos\theta\,\mathrm{d}\varphi\,\mathrm{d}\theta\,, \tag{6.20}$$

and thus from a solid angle Ω

$$J_\Omega = \iint\limits_\Omega L(\varphi,\theta)\sin\theta\cos\theta\,\mathrm{d}\varphi\,\mathrm{d}\theta\,, \tag{6.21}$$

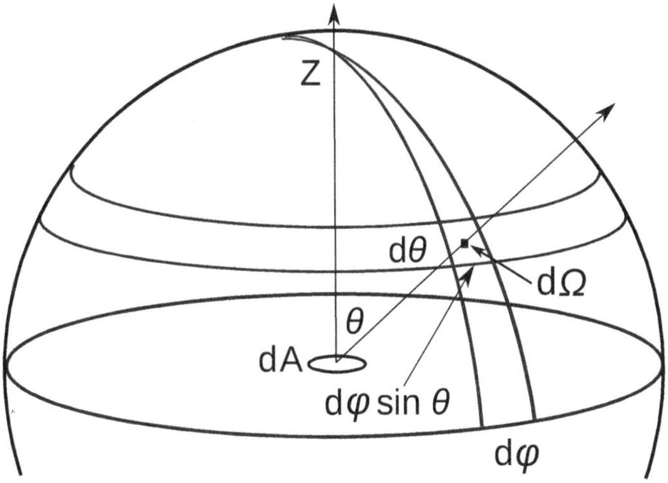

Fig. 6.4: Geometry of a radiation field either emitted (received) from a horizontal surface dA into (from) a solid angle element $d\Omega = \sin\theta d\theta d\varphi$.

and correspondingly from the half sphere above the sensor

$$J_H = \int_0^{\pi/2} \left(\sin\theta\cos\theta \int_0^{2\pi} L(\varphi,\theta)\,d\varphi \right) d\theta \tag{6.22}$$

The radiance integrated over the solid angle of a given source is called *irradiance* in the case of an incoming radiation, and *radiant emittance* in the case of radiation emitted from a radiating surface into space. For an isotropic radiation field with $L(\varphi,\theta) = L_0$, Eq. (6.22) yields

$$J_H = 2\pi L_0 \int_0^{\pi/2} \sin\theta\cos\theta\,d\theta = \pi L_0\,. \tag{6.23}$$

6.3.2 Black-body radiation

Material surfaces emit electromagnetic radiation, called *thermal radiation*, according to their temperature and material properties. In thermal equilibrium, they emit the same electromagnetic energy per unit time as they absorb. A *black body* absorbs all incident radiation and thus constitutes an ideal that is not truly realized in nature. In the visible spectrum, soot is a good approximation of a black body, while it is e.g. ice in the far infrared that is most abundant near room temperature.

Planck radiation law

Real surfaces are never perfect black bodies and although they may absorb a large part of the incident radiation. No material absorbs 100%; the amount of absorption is referred to as the emissivity and depends on the material, the type of surface, the frequency of the radiation and possibly on the direction of the radiation incident on the surface. The emissivity of realistic materials is a very complicated function of the emitted wavelength and its determination may require very laborious and difficult measurements.

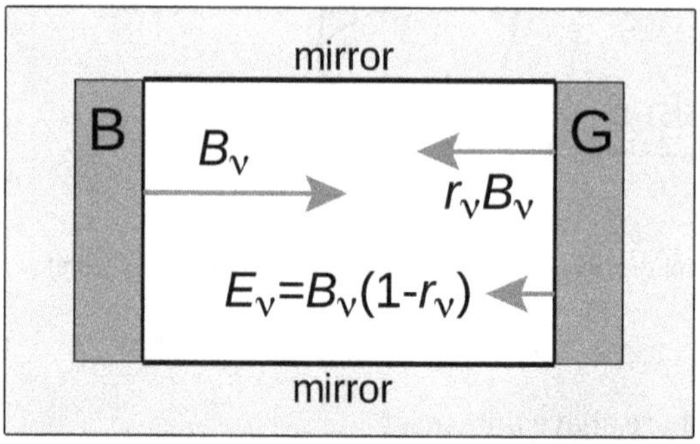

Fig. 6.5: Definition of a grey body: E_ν is the emitted radiation from the grey body G, B_ν the emitted radiation of the black body B, $r_\nu B_\nu$ is the reflected radiation from the grey body and r_ν is the spectral reflectance of the grey body. The mirrors warrant that no radiation between B and G escapes.

Consider the following thought experiment: An ideal black body B and a realistic grey body G are in thermal equilibrium inside a box which is thermally isolated where its inner walls are ideal mirrors (Figure 6.5). Due to the second law of thermodynamics the temperatures of both bodies must be equal.

Between the bodies, the radiation is exactly balanced for all frequencies ν and according wavelengths:

$$B_\nu = E_\nu + r_\nu B_\nu \quad \text{or} \quad \frac{E_\nu}{B_\nu} = 1 - r_\nu \equiv \epsilon_\nu \tag{6.24}$$

where ϵ_ν is the spectral absorptivity (or emissivity) of the grey body (Kirchhoff law). B_ν is the emitted radiation of the black body, E_ν is the emitted radiation from the grey body where $r_\nu B_\nu$ is the reflected radiation from the grey body r_ν is the spectral reflectance of the grey body.

Black-body radiation with a temperature T is described by the *Planck law*,

$$B_\nu \mathrm{d}\nu = \frac{2h\nu^3}{c^2} \frac{\mathrm{d}\nu}{\exp\left(\frac{h\nu}{k_B T}\right) - 1} \tag{6.25}$$

where $B_\nu \mathrm{d}\nu$ is the isotropic spectral radiance [$\mathrm{Wm^{-2}sr^{-1}}$] of the surface of the radiator, ν the frequency, c the speed of light, k_B the Boltzmann constant and h the Planck constant.

Lorentz invariance of the Planck radiation law

If the Planck law is consistent with special relativity, then a black body must remain a black body independent of the state of motion of an observer. This is the case if the radiation temperature is transformed like a frequency (Greber and Blatter, 1990),

$$\frac{T'}{T} = \frac{\nu'}{\nu}\,, \tag{6.26}$$

and the apparent spectral radiance is shifted like the third power of the frequency,

$$\frac{B'_{\nu'}}{B_\nu} = \left(\frac{\nu'}{\nu}\right)^3 . \tag{6.27}$$

To demonstrate this, we analyze the effects of aberration and Doppler shift on the observed radiation of a defined source of monochromatic radiation. Dashed quantities refer to an observer O' with a relative velocity β with respect to a second observer O. The coincident observers measure the powers P and P' of the radiation energy arriving in a given time interval on the sensor area of their radiation detectors,

$$P' = \dot{N}' h\nu' \,; \quad P = \dot{N} h\nu \tag{6.28}$$

where $\dot{N}$ and $\dot{N}'$ are the numbers of arriving photons per unit proper time and sensor area with the frequencies ν and ν', respectively, and h is the Planck constant. Since both, $\dot{N}$ and ν, are frequencies, they appear Doppler shifted, Eq. (2.23),

$$\frac{\dot{N}'}{\dot{N}} = \frac{\nu'}{\nu} = \frac{1}{\gamma(1-\beta\cos\theta')} \equiv q\,, \tag{6.29}$$

and the corresponding energy fluxes on the sensors are transformed like the square of frequencies,

$$\frac{P'}{P} = q^2 . \tag{6.30}$$

Planck radiation is described by an energy flux per solid angle (radiance), thus we need to transform the solid angle Ω of the radiation source. For this transformation, we consider an infinitesimally small circle on the celestial sphere. Since circles are mapped onto circles by aberration (Figure 6.6),

$$\frac{\mathrm{d}r'}{\mathrm{d}r} = \frac{\mathrm{d}\theta'}{\mathrm{d}\theta} = \frac{\sin\theta'}{\sin\theta}\,, \tag{6.31}$$

and the solid angle of the circle transforms like the square of its angular radius,

$$\frac{\mathrm{d}\Omega'}{\mathrm{d}\Omega} = \left(\frac{\mathrm{d}r'}{\mathrm{d}r}\right)^2 = \frac{\mathrm{d}\theta'}{\mathrm{d}\theta}\,\frac{\sin\theta'}{\sin\theta} \tag{6.32}$$

Differentiating the second Eq. (2.17) with respect to θ and substitution in Eq. (6.32) yields

$$\frac{\mathrm{d}\Omega'}{\mathrm{d}\Omega} = \frac{1}{q^2} . \tag{6.33}$$

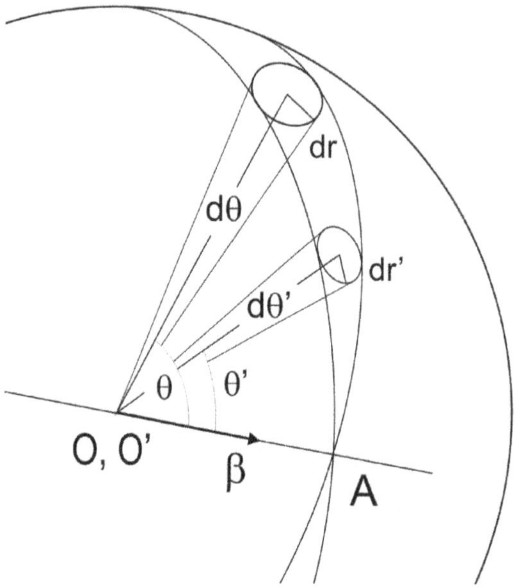

Fig. 6.6: Transformation of the solid angle of an infinitesimal circle due to aberration (see Figure2.3).

Finally, the spectral radiance $B = I/\mathrm{d}\Omega$ of the radiation source transforms like the fourth power of a frequency,

$$\frac{B'}{B} = q^4 . \tag{6.34}$$

For a given spectral distribution of the radiation of an emitting source, this transformation applies to any infinitesimally small frequency range, and thus, it also applies for the integral over a given range of frequencies, if the limits of the ranges coincide for the observers according to the Doppler shift.

For the integral of the black-body radiation over the whole spectrum, this yields the Stefan-Boltzmann law,

$$I_{\mathrm{tot}} = \pi \int_0^{\infty} B_\nu \mathrm{d}\nu = \sigma T^4 , \tag{6.35}$$

where $\sigma = 5.670374419 \cdot 10^{-8}\ \mathrm{Wm^{-2}K^{-4}}$ is the Stefan-Boltzmann constant that depends on k_B, h and c. The factor π stems from the integration of the 4π solid angle and the $\cos\theta$ dependence, Eq. (6.23). With Eq. (6.34), the required transformation of the radiation temperature, Eq. (6.26) is met. The application of transformations (6.26), (6.33) and (6.34) to the Planck law, Eq. (6.25), yields

$$B'_{\nu'}\,\mathrm{d}\nu' = \frac{2h\nu'^3}{c^2}\,\frac{\mathrm{d}\nu'}{\exp\left(\frac{h\nu'}{kT'}\right) - 1} \tag{6.36}$$

and thus, Eq. (6.27) is also fulfilled.

6.3.3 Accelerated charges

A charged particle moves with constant velocity along a straight line as long as there are no external electromagnetic nor gravitational fields along the trajectory.

It is a known fact that accelerated electric charges emit electromagnetic waves. Here, accelerated is used in the sense of Newton's second law, i.e. that any charge on which acts a force, changes its state of motion and emits electromagnetic radiation. Still, we omit the discussion of the atoms, where the orbiting electrons are accelerated, though where states of motion are observed without emission of radiation (see Section 1.4.4).

In the rest frame of the charge, the emission has cylindrical symmetry along the acceleration axis with $\sin^2\vartheta$ characteristics, where ϑ is the angle measured from the acceleration direction (see Figure 6.7).

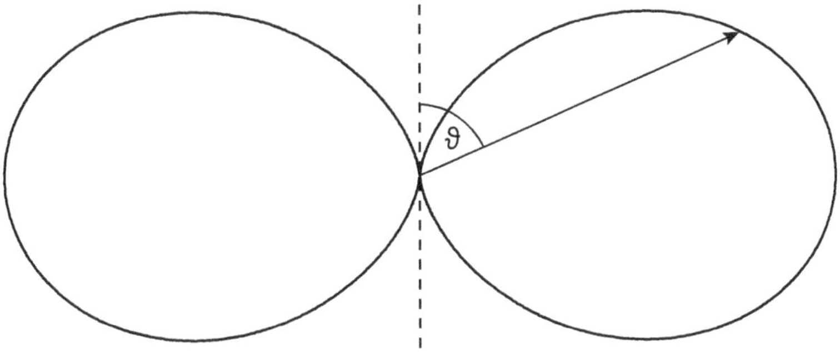

Fig. 6.7: Angular distribution of the average emitted power of an oscillating dipole in the far field. The radiant power intensity $I(\vartheta)$ is cylindrically symmetric with respect to the acceleration direction of the charge (dashed line), and has $\sin^2\vartheta$ characteristics.

The characteristics in the rest frame of the source are the same for linear acceleration, circular motion, or linear deceleration as long as the absolute value of the force is the same. The node along the acceleration direction may be understood by the nature of the emitted energy in the form of electromagnetic radiation, where the electric field along the acceleration axis is perpendicular to the magnetic field, Section (3.3.1). In a moving frame of reference, the detected emission characteristics will furthermore depend on the orientation of the velocity, relative to the acceleration vector, and we have to distinguish linear acceleration and perpendicular acceleration with respect to the velocity vector. In the following, we will use θ and θ' as the angles measuring directions with respect to the velocity vector, where ϑ is the angle with respect to the acceleration direction in the rest system of the radiation source.

Charged particles that are stopped rapidly emit bremsstrahlung i.e. a pulse of x-rays, charged particles moving in a circle emit so-called synchrotron radiation or oscillating charge emits Hertz dipole radiation.

Hertz dipole

An oscillating electric charge distribution emits electromagnetic radiation. The simplest example is a point charge oscillating harmonically in one dimension and forming a time-dependent dipole. The radiation of such an oscillating charge has cylindrical symmetry with respect to the oscillation axis, also called dipole axis. The description of radiation fields of accelerated charges requires mathematical treatment beyond this book. Still, we give the relevant results that are applied for description of collimation due to aberration and Doppler effect.

The radiant power intensity [W sr^{-1}] of an accelerated charge such as an oscillating dipole in the far field is given by (Lorrain et al., 1988; Jackson, 1975),

$$\frac{dP}{d\Omega} = I(\vartheta) = I_0 \sin^2 \vartheta \,, \tag{6.37}$$

where ϑ is the angle between the axis of the dipole and the emission direction in the rest frame of the accelerated charge (Figure 6.7). I_0 is a coefficient depending on the power of the emitter. Far field conditions are obtained if the oscillation amplitude is much smaller than the distance to the oscillator. The radiant power intensity I is measured in energy per time and solid angle. To obtain the total emitted radiant power P, we integrate Eq. (6.37) over the entire celestial sphere. For this purpose, the dipole axis is chosen parallel to the z-axis in Figure 6.4 and $\vartheta = \theta$, thus

$$P = \int_0^{\pi}\int_0^{2\pi} I(\theta)\, \sin\theta \,\mathrm{d}\varphi \,\mathrm{d}\theta = 2\pi I_0 \int_0^{\pi} \sin^3\theta \,\mathrm{d}\theta = \frac{8\pi}{3} I_0 \,. \tag{6.38}$$

I_0 is proportional to the square of the oscillating charge and proportional to the square of the acceleration. In the case of variable acceleration, as it is given for the Hertz dipole, the mean power emitted is proportional to the square of the acceleration averaged over the time interval in question, $\langle a^2 \rangle$ (Lorrain et al., 1988; Jackson, 1975).

Given a harmonic oscillation of a charge q with amplitude A, angular frequency w, location s, velocity v and acceleration a as a function of time are described by

$$s = A \sin(wt), \tag{6.39}$$

$$v = \dot{s} = Aw \cos(wt), \tag{6.40}$$

$$a = \ddot{s} = -Aw^2 \sin(wt)\,, \tag{6.41}$$

and the average square of the acceleration is

$$\langle a^2 \rangle = (A\,w^2)^2 \,\frac{\int_0^{2\pi/w} \sin^2(wt)\,\mathrm{d}t}{\int_0^{2\pi/w} \mathrm{d}t} = \frac{1}{2} A^2\, w^4\,, \tag{6.42}$$

thus, the emitted power is proportional to the fourth power of the frequency and proportional to the square of the amplitude A or the oscillating dipole $p = Aq$. Quantitatively the power coefficient I_0 for the Hertz dipole gets:

$$I_0 = \frac{p^2\, w^4}{32\pi^2\, \epsilon_0\, c^3}\,. \tag{6.43}$$

Notably, the $\sin^2\vartheta$ emission characteristics (Eq. (6.37), and Figure 6.7) are also valid for non-oscillating, but unidirectionally accelerated charges.

Relativistic collimation

In the frame of reference of the dipole, the radiation field has cylinder symmetry as shown in Figure 6.7. In a system where the dipole is moving, aberration collimates the radiation in the direction of the apex of the motion.

We consider two possible geometries for the orientation of the acceleration or dipole vector $\mathbf{p}$ or $\dot{\vec{\beta}}$ and the relative velocity β: *(i)* the bremsstrahlung geometry where $\mathbf{p}$ is parallel to β, and *(ii)* the synchrotron geometry, where $\mathbf{p}$ is perpendicular to β (see Figure 6.8).

(i) Bremsstrahlung (longitudinal) geometry
Let the dipole move with the primed system, then

$$\frac{\mathrm{d}P'}{\mathrm{d}\Omega'} = I_0\, \sin^2\theta'\,. \tag{6.44}$$

For an observer in the unprimed system, due to aberration, Eq. (2.16)

$$\sin^2\theta' = \frac{\sin^2\theta}{\gamma^2\,(1-\cos\theta)^2}\,, \tag{6.45}$$

due to the Doppler shift and time dilation, Eqs. (2.23) and (4.46),

$$\mathrm{d}P' = I_0\, \frac{1}{\gamma^2\,(1-\cos\theta)}\,, \tag{6.46}$$

and with the transformation of the solid angle due to aberration, Eq. (6.33),

$$\mathrm{d}\Omega' = \gamma^2\,(1-\cos\theta)^2\,\mathrm{d}\Omega\,. \tag{6.47}$$

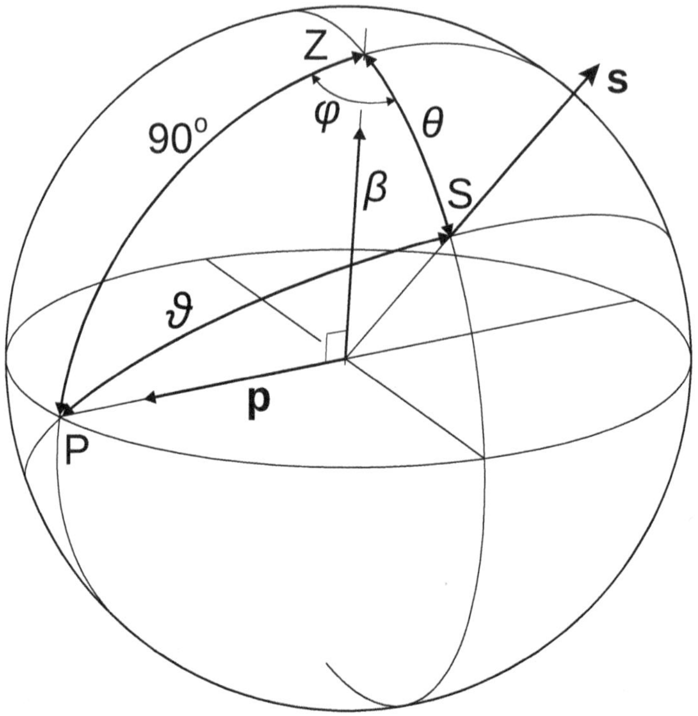

Fig. 6.8: Geometrical setup for a moving Hertz dipole in spherical coordinates θ and φ. The velocity β is orthogonal to the axis of the dipole $\mathbf{p}$ (synchrotron direction). **s** is the viewing direction.

With this the Bremsstrahlung (dipole collimation) finally becomes

$$I = \frac{dP}{d\Omega} = \frac{I_0 \sin^2\theta}{\gamma^6 (1 - \beta\cos\theta)^5}, \tag{6.48}$$

where θ is the polar angle measured from the apex of β. A cross-section of the radiation distribution for this motion is shown in Figure 6.9 (a). The intensity distribution has cylinder symmetry and it can be seen that the aberration bends and amplifies the two $\sin^2\theta'$ lobes towards the apex. At certain solid angles, the radiant power intensity I may exceed that of the rest system of the emitter. Along $\beta = 0$ I is zero. For highly relativistic cases ($\gamma \gg 1$) the maximum radiant power intensity is

$$I_{\max} = \frac{8192}{3125}\gamma^2 I_0 \quad \text{at} \quad \theta_{\max} = \frac{1}{2\gamma}. \tag{6.49}$$

The total emitted power is independent of the velocity β. This can be shown by integrating Eq. (6.48) over the entire sphere, similar to Eq. (6.42),

$$\begin{aligned} P &= \int_0^{\pi}\int_0^{2\pi} \frac{I_0 \sin^3\theta}{\gamma^6 (1-\beta\cos\theta)^5}\, \mathrm{d}\varphi\, \mathrm{d}\theta \\ &= 2\pi \frac{I_0}{\gamma^6} \frac{1 - 4\cos\theta + 3\beta^2 \cos(2\theta)}{12\beta^3 (1-\beta\cos\theta)^4}\Bigg|_0^{\pi} = \frac{8\pi}{3} I_0 \equiv P\,. \end{aligned} \tag{6.50}$$

(ii) Synchrotron (transversal) geometry

For the synchrotron (transversal) geometry the velocity β is orthogonal to the axis $\mathbf{p}$ of the dipole (see Figure 6.8).

With the spherical law of cosines (spherical triangle PSZ in Figure 6.8) we obtain

$$\frac{\mathrm{d}P'}{\mathrm{d}\Omega'} = I_0 \sin^2\Theta' = I_0 (1 - \cos^2\varphi' \sin^2\theta') \tag{6.51}$$

With the same procedure as for the Bremsstrahlung, Eq. (6.48), Eqs. (6.45), (6.46) and (6.47), we obtain

$$I = \frac{\mathrm{d}P}{\mathrm{d}\Omega} = \frac{I_0}{\gamma^4 (1-\beta\cos\theta)^3}\left[1 - \frac{\sin^2\theta\cos^2\varphi}{\gamma^2 (1-\beta\cos\theta)^2}\right]. \tag{6.52}$$

The angle φ is not affected by the movement. As for the Bremsstrahlung, the total emitted power of the synchrotron radiation is independent of the velocity β.

For the plane spanned by β and $\mathbf{p}$ Figure 6.9 (b) we get

$$I = \frac{I_0}{\gamma^4 (1-\beta\cos\theta)^3}\left[1 - \frac{\sin^2\theta}{\gamma^2 (1-\beta\cos\theta)^2}\right]. \tag{6.53}$$

The opening angle θ_c of the collimated dipole radiation is defined by the apparent direction of the dipole axis, $\theta_c = \pi/2$, and thus with Eq. (2.16),

$$\sin\theta_c = \sqrt{1-\beta^2} = \frac{1}{\gamma}, \tag{6.54}$$

which for highly relativistic velocities can be approximated by $\theta_c = 1/\gamma$.

For the plane spanned by β and the direction perpendicular to β and $\mathbf{p}$ Figure 6.9 (c) we get

$$I = \frac{I_0}{\gamma^4 (1-\beta\cos\theta)^3}. \tag{6.55}$$

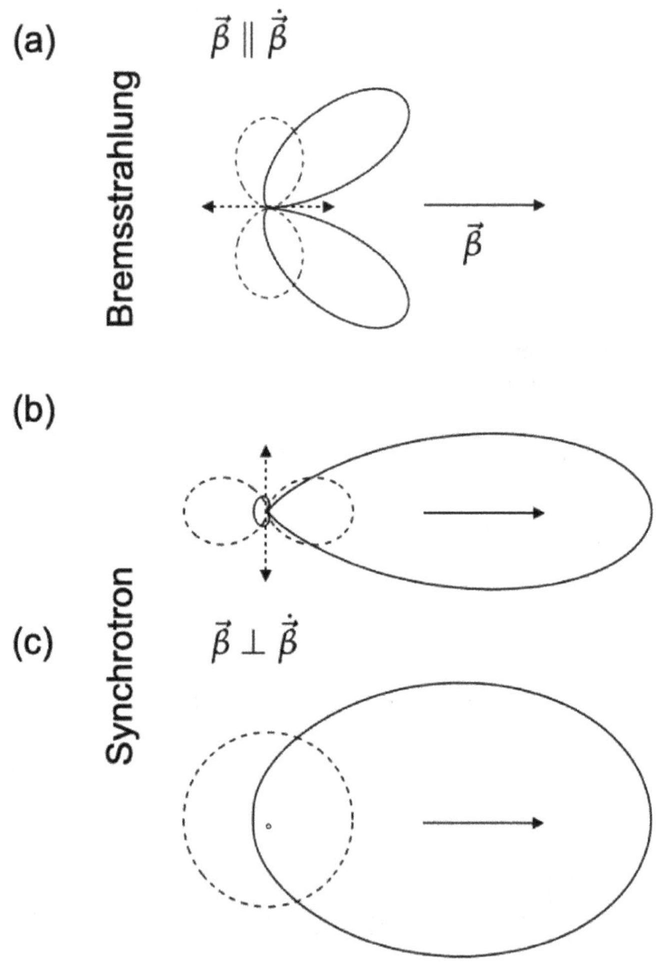

Fig. 6.9: Angular distribution of the radiation power $dP/d\Omega$ of a Hertz dipole **p** with the axis along $\dot{\beta}$. The dashed lines represent the $\sin^2 \vartheta$ characteristics in the rest system of the emitter, the solid lines that of the emitter in a moving frame with velocity $\beta = 0.5$. (a) Bremsstrahlung geometry $\beta \parallel \dot{\beta}$. (b,c) Synchrotron geometry $\beta \perp \dot{\beta}$.

Relativistic spectrum

In the rest frame of the Hertz dipole, the spectrum of the radiation is monochromatic. The Doppler effect shifts the frequency of the radiation, in a frame of reference that moves relative to the Hertz dipole. For the synchrotron geometry scenario in Figure 6.9 and Eq. (2.23) a frequency enhancement factor along the apex of $\approx 2\gamma$ is expected, while along the antapex the spectrum is red-shifted accordingly.

Particle acceleration

Particles with high velocities are used in basic science on elementary particles and nuclear physics or to create intensive beams of electromagnetic waves again for a variety of scientific investigations e.g. in surface physics, chemistry and material technology, but also in applications such as medical treatment of cancer and the generation of x-rays for the detection of anomalies in the human body. Depending on the application, the particles may be electrons, protons, or heavier ions, and may span a large range of energies.

For the acceleration, various methods are used, from simple acceleration of the particles due to the Coulomb force in a linear electrostatic field (Van-de-Graff accelerators) to linear particle accelerators using radio waves to boost particles in a resonant mode (LINACS), or by repeated boosting with electric fields while orbiting in circles in a homogeneous magnetic field and repeated acceleration by electric fields on orbital cycles (cyclotrons).

Particle storage, Synchrotron radiation

Storing particles at relativistic energies can be obtained by keeping them in circular orbits, where the acceleration of their charge leads to so-called synchrotron light emission.

For highly relativistic velocities, the diameter of the circular orbit becomes too large for a single dipole magnet as it is used in cyclotrons. For this reason, many smaller bending magnets are aligned along the path. Between the magnets, straight sections may be used for the acceleration of the particles in order to compensate for the energy loss due to the emitted radiation and to refocus the beam. Similar to linear accelerators, in synchrotrons particles are accelerated in batches with phase-controlled radiation, with the microwave frequency corresponding to the orbit period. The emitted electromagnetic radiation (see Section 6.3.3) makes a synchrotron a source of electromagnetic radiation from ultraviolet to hard x-ray frequencies.

The Swiss light source (SLS) near the village of Villigen (see Figure 6.10) is an x-ray source with extremely high brightness. Brightness is the radiant power intensity per frequency interval $d\nu$. The facility consists of three major parts, first a *linear accelerator (Linac)*, secondly a *synchrotron* and thirdly a *storage ring* for high energy electrons. In the Linac, a batch of electrons is accelerated through two 5.2 m long linear accelerators to an energy of 100 MeV. From the Linac, the electrons are injected into the *synchrotron* which accelerates the electrons to a final energy of 2.4 GeV, corresponding to $\gamma = 4700$ or a velocity of $\beta = 0.999999955 = (1\text{-}4.5 \cdot 10^{-8})$ into the storage ring with 288 m circumference. The boosting and injection process is repeated three times per second until after a few seconds, the storage ring is filled with several trillions of electrons orbiting the ring. Due to the synchrotron radiation, an electron in such an orbit loses 25 ppm of its energy in one revolution, which lasts about 1 μs in the rest frame of the SLS .

Fig. 6.10: The Swiss Light Source synchrotron in Villigen is housed in the largest wooden building in Switzerland with an outer diameter of 138 m and a height of 14 m. Foto: M. Fischer, Paul Scherrer Institut, 2014.

Bending magnets

The simplest type of device to generate light is a *bending magnet* where the electrons are deflected in a circular orbit. Every deflected electron emits flashes of synchrotron radiation in a picture reminiscent to a "lighthouse", though with MHz repetition rates. In the frame of the laboratory, this light is strongly collimated in the direction of the motion of the electrons due to aberration and Doppler shift. For a given kinetic energy the orbital velocity and consequently the centripetal acceleration are larger for light particles, such as electrons, than for heavy particles, such as protons or ions. Therefore, electrons radiate more light than heavy particles with the same kinetic energy, which is the reason why electrons or positrons are used in synchrotrons as x-ray sources.

The directional distribution of the synchrotron radiation on a circular path is collimated in the direction of the apex of the motion, i.e. tangential to the orbit. The directions parallel to the plane normal to the velocity vector are projected into directions along a cone of total opening angle $\alpha \approx 2/\gamma$, Eq. (6.54). In the laboratory frame, half of the emitted energy is collimated into this cone. The radiance transforms with the fourth power of the transformation factor of the frequency, Eq. (6.34).

In the following we consider a light beam which covers an angle $2/\gamma$. Figure 6.11 illustrates how the duration Δt of the light flash of one particle is determined, as it is seen by an observer a distance R away from the electron orbit. Two emission points of photons from an electron on the circular orbit with radius r are separated by the angle $2/\gamma$. They represent the leading and the trailing edge of the light beam, as it is the beam

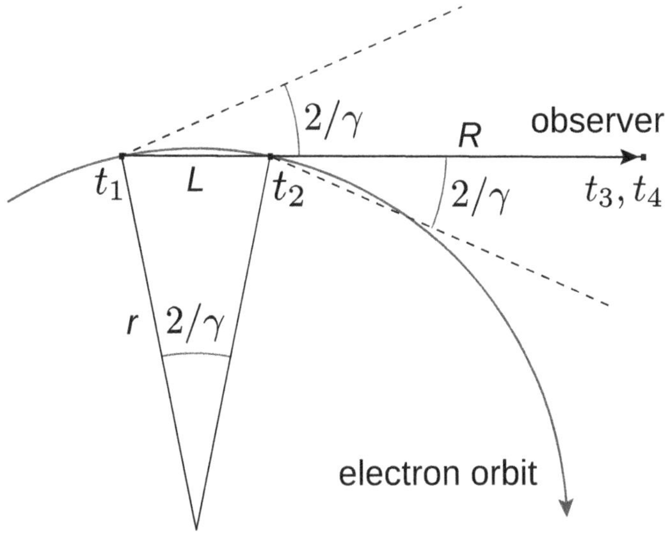

Fig. 6.11: Schematic of the "lighthouse effect" from which the duration of the light flashes may be calculated. An electron on orbit r emits photons in a ray with a divergence angle of $1/\gamma$ towards an observer a distance R away from the orbit. Light emitted from the leading edge at t_1 arrives at t_3 at the observer. Light emitted from the trailing edge at t_2 arrives at t_4 at the observer.

emitted from a lighthouse. The emission times are t_1 and t_2, where their emission sites are separated by $L = 2r/\gamma$ (Figure 6.11). The emission times differ by

$$t_2 - t_1 = \frac{L}{c\beta} = \frac{2r}{c\beta\gamma} . \tag{6.56}$$

The emitted light therefore arrives at the observer at times t_3 and t_4, thus a flash of duration

$$\Delta t = t_4 - t_3 = t_2 + \frac{R}{c} - t_1 - \frac{L+R}{c} , \tag{6.57}$$

which is in the limit of for large γ, thus $\beta \to 1$,

$$\Delta t = \frac{r}{c\gamma^3} . \tag{6.58}$$

The radius of the SLS bending magnet trajectories is $r = 5.729$ m. Thus, an observer (Figure 6.11) sees flashes of $\Delta t \approx 1.8 \cdot 10^{-19}$ s duration emitted by each electron. With a spectral Fourier analysis, we find an upper limit of the possible frequency for this kind of pulsed signals of

$$\nu_{max} \approx \frac{1}{\Delta t} = \frac{c\gamma^3}{r} , \tag{6.59}$$

and obtain an upper cut-off frequency of $5.4 \cdot 10^{18}$ Hz corresponding to light with 56 pm wavelength, or a photon energy of 22 keV, which is in the range of x-rays useful for atomic structure determination.

Particle stopping

A charged particle that is stopped emits electromagnetic radiation. This may happen in an electric field generated by some electronic device like a radio antenna or if a charged particle is stopped in a target.

This radiation due to stopping of charged particles was discovered by Röntgen (1895) These "cathode rays" were called x-radiation or Roentgen radiation, and were identified to be electromagnetic radiation by C.G. Barkla who also discovered "secondary" x-radiation that is produced by x-ray fluorescence of electrons in atoms, Barkla (1903).

As both, charged particles with high kinetic energy, and x-rays penetrate matter they can be used to look inside material objects, or to treat them below the surface. Compared to x-rays, charged particles have a completely different energy dissipation pattern. This makes them complementary and more or less versatile for a given application. High-energy particles are also used in medical treatment of cancer and tumors. Like x-rays, the ionizing effect of the interaction of highly energetic charged particles and matter also destroys the tissue. The general problem of radiation therapy that it also destroys healthy tissue may be smaller with high energy particles since they dissipate their energy differently. With charged particles cancer tissue at a known depth may be destroyed with minimal damage of healthy tissue: the largest part of the energy of the particles is deposited in a narrow range of the tissue shortly before the particles come to a halt.

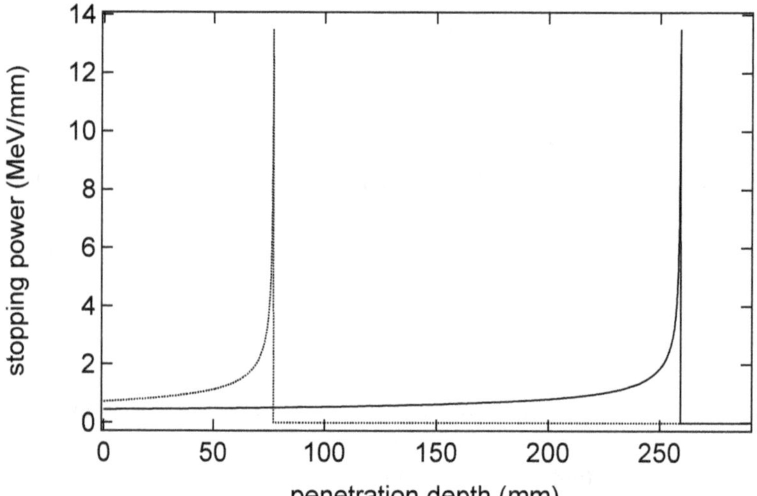

Fig. 6.12: Stopping power or deposition of kinetic energy per path length as a function of penetration depth, dE/dz, of protons with 100 MeV (dotted line) and 200 MeV (solid line) kinetic energy in water. The continuous slowing down approximation (CSDA) predicts the stopping depths of 77 and 260 mm before which the Bragg peaks occur. The full width at half maximum of the Bragg peaks is less than 1 mm. Data from tables of the national institute of standards and technology (NIST).

If a particle is stopped in a material, the stopping power or dissipated energy $\mathrm{d}E$ per step $\mathrm{d}z$ depends on the corresponding scattering cross-section of the particle. The maximum dissipation occurs just before the particle comes to rest. This peak of dissipation is called *Bragg peak*, Wilson (1946). The energy deposition after the peak is zero and before the peak it is smaller. Thus, the particles cross the tissue with less damage before they dissipate most of their energy in a desired spot.

For the case of protons with 150 (230) MeV kinetic energy it is experimentally found that they stop in water, which is a reasonable model for tissue, at a depth of about 140 (300) mm Kumazaki et al. (2007). For a proton with a rest energy $m_p c^2$ of 938 MeV this corresponds with Eq. (3.44) to coordinate velocities β of 0.5 (0.6), respectively. Figure 6.12 shows the expected energy dissipation of high energy protons in water. Next we will compare the particle stopping in a non-relativistic and a relativistic picture.

Non-relativistic case

The interaction of the particles with matter includes deflection and transfer of energy quanta. The stopping force acting on the particles can be parametrized. While drag in turbulent media scales with v^2 and in laminar flow with v. Here we explore friction that scales $\sim v^{-1}$. Intuitively this is motivated in a picture where the dissipated power scales with the time $\mathrm{d}t$ a particle spends on a path element $\mathrm{d}z$. In a one-dimensional model for a single particle, where the randomness of the energy loss is neglected, the deceleration is

$$a = -\frac{\mathrm{d}v}{\mathrm{d}t} = -\frac{A}{v}\,, \tag{6.60}$$

where the sign indicates deceleration and A is the proportionality factor that e.g. depends on the density of the stopping medium and the stopping particle. With $\mathrm{d}t = \mathrm{d}z/v$ we obtain

$$v\,\mathrm{d}v = -A\,\mathrm{d}t = -A\,\frac{\mathrm{d}z}{v}\,, \tag{6.61}$$

which is equivalent to the differential equation for $v(z)$, separated by variables:

$$v^2\,\mathrm{d}v = -A\,\mathrm{d}z\,. \tag{6.62}$$

This equation is solved by integration,

$$v^3 - v_0^3 = -3A\,z \tag{6.63}$$

where v_0 is the velocity at $z = 0$. Thus the velocity $v(z)$ is

$$v(z) = \sqrt[3]{v_0^3 - 3A\,z} \tag{6.64}$$

and accordingly, the kinetic energy as a function of the depth z is

$$E_{\mathrm{kin}}(z) = \frac{1}{2}\,m\,v^2(z) = \frac{1}{2}m\,\sqrt[3]{v_0^3 - 3A\,z}^{\,2}\,. \tag{6.65}$$

The dissipated energy per unit path length dz becomes

$$\frac{dE_{\text{kin}}}{dz} = -\frac{m\,A}{\sqrt[3]{v_0^3 - 3A\,z}}\,. \tag{6.66}$$

When the root in Eq. (6.66) equals 0, the particle is at rest at the stopping depth

$$z_s = \frac{v_0^3}{3A}\,. \tag{6.67}$$

Figure 6.13 compares the expected kinetic energy of a 200 MeV proton with the non-relativistic friction model of Eq. (6.60). With Eq. (6.65) and Eq. (6.67) the non-relativistic $1/v$ friction model for $z \le z_s$ gets:

$$E_{\text{kin}}(z) = E_0\left(1 - \frac{z}{z_s}\right)^{\frac{2}{3}}, \tag{6.68}$$

where E_0 is the initial kinetic energy. The model qualitatively describes $E_{\text{kin}}(z)$, though the relativistic treatment as outlined below does not reconcile the expected kinetic energy dissipation, which is due the fact that the parameter A in Eq. (6.60) still depends on the kinetic energy. In brackets we note that if the exponent 2/3 in Eq. (6.68) is left as an empirical parameter we obtain with 0.567 an almost perfect fit to the expected kinetic energy dissipation of 150 MeV protons in water.

Relativistic case

The relativistic 2nd law, Eq. (3.24) relates the proper velocity ω proportional to the acting force. Similar to Eq. (6.60) we write for the relativistic deceleration

$$a_r = c\,a = -c\,\frac{d\omega}{dt} = -\frac{A}{\omega\,c} \tag{6.69}$$

Separation of variables yields

$$c^2\,\omega\,d\omega = -A\,dt = -A\,\frac{\gamma}{\omega\,c}\,dz\,, \tag{6.70}$$

written in ω and z this yields

$$c^3\,\frac{\omega^2}{\sqrt{1+\omega^2}}\,d\omega = -A\,dz\,. \tag{6.71}$$

Integration of Eq. (6.71) yields

$$z = z_{sr} - \frac{c^3}{2A}\left[\omega\,\sqrt{1+\omega^2} - \ln\left(\omega + \sqrt{1+\omega^2}\right)\right], \tag{6.72}$$

where the integration constant

$$z_{sr} = \frac{c^3}{2A}\left[\omega_0\,\sqrt{1+\omega_0^2} - \ln\left(\omega_0 + \sqrt{1+\omega_0^2}\right)\right] \tag{6.73}$$

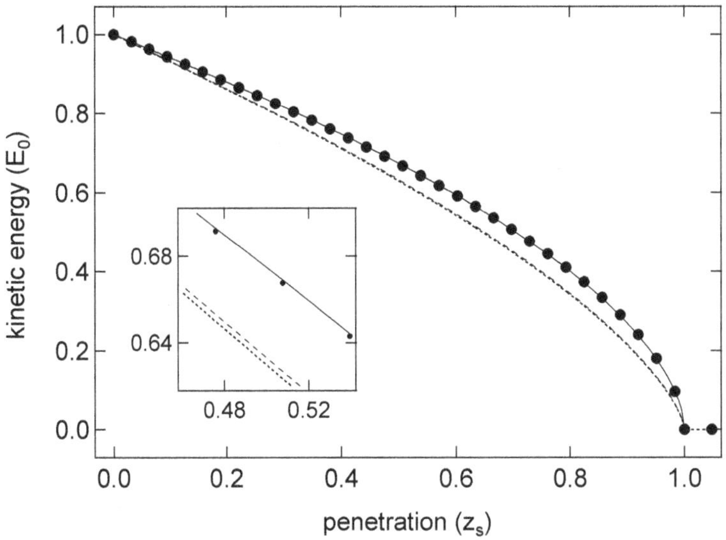

Fig. 6.13: Energy of a charged particle during penetration in water. Dots: NIST data for E_0=150 MeV protons in water, z_s= 158 mm. Dotted line: non-relativistic $1/v$ model Eq. (6.60). Dashed line: relativistic $1/\omega$ model Eq. (6.69). Solid line: exponent 2/3 in Eq. (6.68) replaced with 0.567. The inset exhibits the differences between the different descriptions. The relativistic description has a small influence, only.

is the relativistic stopping length. With this the deposition of energy per length gets:

$$\frac{\mathrm{d}E_{\mathrm{kin}}}{\mathrm{d}z} = \frac{\mathrm{d}E_{\mathrm{kin}}}{\mathrm{d}\omega}\frac{\mathrm{d}\omega}{\mathrm{d}z}, \tag{6.74}$$

where the derivatives can be obtained from Eq. (3.39) and Eq. (6.71),

$$\frac{\mathrm{d}E_{\mathrm{kin}}}{\mathrm{d}z} = -\frac{A\,m\,c^2}{\omega}, \tag{6.75}$$

which is a remarkably simple expression.

It is seen that stopping of high energy protons as they are used in proton therapy does not alter significantly if the relativistic force law is applied. This is not the case in hyperrelativistic space travel as discussed in Chapter 7.

6.4 Relativistic effects in atoms

The Dirac equation describes particles like electrons or positrons in spacetime (Dirac, 1928). It implies concepts of spin and antimatter and makes predictions on the wave functions of the particles. In hydrogen atoms, the electrons move relative to the protons with velocities in the order of $c/137$, where $1/137$ is the fine structure constant α (see Section 1.4.4). Therefore relativistic effects can be treated as weak perturbations and for most applications the non-relativistic Schrödinger equation that describes particles as waves is appropriate. The solutions of the Schrödinger equation are as well wavefunctions, where the square of the amplitude describe for the case of the hydrogen atom the probability to find the electron at a given place in space. High-precision microwave spectroscopies allow the test of relativistic effects to unprecedented precision, and, the spin of the electrons moving relative to the atomic nuclei yields magnetism, which is an essential quantum phenomenon with strong implications for technological possibilities.

6.4.1 The ionization energy of hydrogen

Like the Bohr model of the hydrogen atom that describes the electron as a particle that orbits the proton, the Schödinger equation predicts the ionisation energy of the hydrogen atom to be equal to the Rydberg energy E_R:

$$E_R = \frac{\alpha}{4\pi a_o} h\, c\,, \tag{6.76}$$

where α is the fine structure constant, Eq. (1.10), and a_o the Bohr radius, Eq. (1.9), h the Planck constant and c the speed of light. The true ionization energy of hydrogen deviates from $E_R \simeq 13.6$ eV because of the finite mass of the proton, the velocity of the electron, the deviation of the electron and the proton from a point, and the spins of the electron and the proton. Besides the effect of the finite mass of the proton, the description of these effects involves the Dirac equation and deeper quantum electrodynamics. Here we describe the effects as a perturbation of the non-relativistic electron 1s wave function from the Schrödinger equation that has the lowest possible energy. In Figure 6.14 the hydrogen ground state energies for different theory levels are depicted. As a convention the ground state energy is negative and the ionization energy has to be added in order to separate the electron from the proton that results in a state with zero kinetic and zero potential energy.

The finite mass of the proton calls for the use of the reduced electron mass for the description of the dynamics in the hydrogen atom. The reduced mass is

$$\mu_r = \frac{m_p\, m_e}{m_p + m_e} = \frac{1836}{1837}\, m_e\,, \tag{6.77}$$

where m_p and m_e are the proton and electron mass, respectively. It is a non-relativistic effect and reduces the ionization energy of hydrogen atoms by a factor of $(1\text{-}5.4 \cdot 10^{-4})$.

The velocity of the electron imposes a relativistic term in the kinetic energy, Eq. (3.39), which affects the ionization energy. Adopting the Bohr model, and the first order expansion of the relativistic kinetic energy $\frac{1}{2}mv^2(1-\frac{1}{4}\beta^2)$ we get for the hydrogen atom an increase of the ionization energy by a factor of $1+3/4\alpha^2$ or about $(1+6.6 \cdot 10^{-5})$.

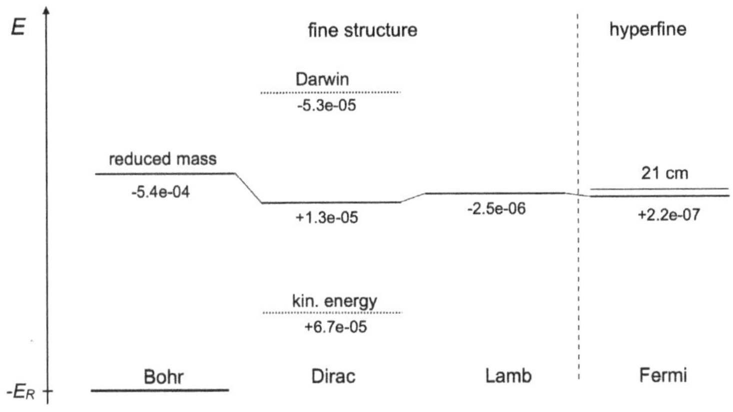

Fig. 6.14: The ground state energy of the hydrogen atom from different levels of theory. The ground state of $-E_R \simeq$-13.6 eV is obtained from the Bohr model. Fine structure: the most significant correction by a factor of 1-5.4 $\cdot 10^{-4}$ is due to the reduced mass. Relativistic corrections as described by the Dirac theory lead to the kinetic energy term and the Darwin term that almost cancel to 1+1.3 $\cdot 10^{-5}$. The Lamb shift is a further correction decreasing the ionization energy. Magnetic hyperfine interaction between the spin of the electron and the spin of the proton leads to the weakest correction that was first calculated by Fermi (1930).

The extension of the electron from a point to the Compton wavelength

$$\lambda_C = \hbar / m_e c = \alpha a_0 , \tag{6.78}$$

causes a decrease of the ionization energy due to the smaller overlap of the ground state 1s wave function with the extended proton that yields less attraction of the electron. This is the Darwin term that amounts to a factor of $1 - \alpha^2$ or about (1-5.31 $\cdot 10^{-5}$).

Both, the velocity and the extension of the electron are so-called Dirac terms that are predicted by the Dirac equation and that scale with α^2. A further correction that was first recognized in a small energy difference between the hydrogen 2s and the hydrogen 2p states by Lamb and Retherford (1947) is not predicted by the Dirac equation. Like the Darwin term, this Lamb shift is limited to s levels. It has the same sign, though a smaller magnitude and may be rationalized again with a finite radius of the proton and is about (1-2.5 $\cdot 10^{-6}$).

Finally, we mention the hyperfine structure of hydrogen, which is due to the magnetic interaction of the proton spin and the electron spin. The transition between the two states with parallel or antiparallel spins has a wavelength of 21 cm and was first calculated by Fermi (1930). In radioastronomy it serves for the detection of interstellar hydrogen (Ewen and Purcell, 1951). The importance of the 21 cm hydrogen line is also documented as an engraving on the plaquette of the pioneer space probes designated to leave our solar system. This hyperfine interaction is the weakest correction of the ionization energy of hydrogen, which would increase it for the lower-lying state by a factor of $(1 + 2.2 \cdot 10^{-7})$.

6.4.2 Spin-orbit interaction

The spin-orbit interaction causes the corresponding spin-orbit splitting in atomic spectra. It is a relativistic effect in atoms and a key ingredient for the understanding of magnetism. It arises due to the interaction of the electron spin magnetic moment and the B-field in the rest system of the electron due to the Lorentz transformation of the electric field of the nucleus. If the electron has an angular momentum, a net energy shift is observed between the up and the down spin configuration of the electron (Section 1.4.3).

Most free atoms have an open electron shell with a net angular momentum of the electrons. Boron is the lightest atom with a net electron angular momentum in its ground state. Here the electron, described by the $2p$ wave function, carries a quantized angular momentum of $h/2\pi$. The spin vector of the electron may be either parallel or antiparallel to the angular momentum vector of the $2p$ electron. Like in the experiment of Gerlach and Stern (1922) the two spin states couple differently to a parallel or antiparallel external magnetic field. For the case of the boron $2p$ electron, a magnetic field is *intrinsically* provided, via the Lorentz transformation of the Coulomb field of the nucleus. In a **B**-field

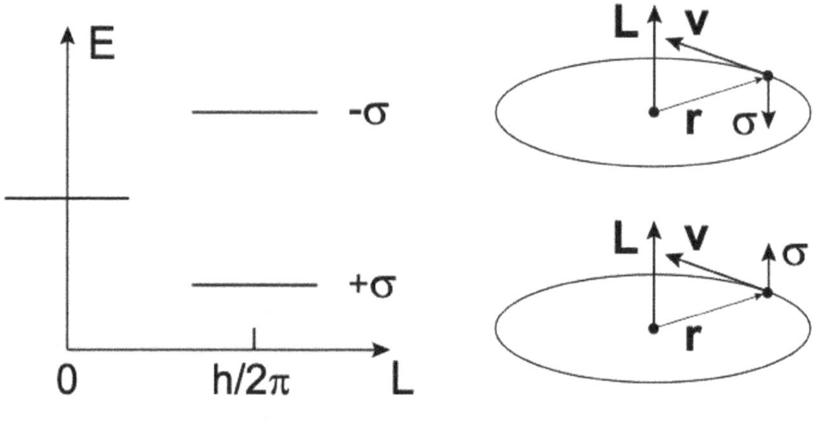

Fig. 6.15: The energy of an electron on an orbit with angular momentum **L** can take two values, depending on whether the spin σ is parallel + or antiparallel − to **L**. This effect of the spin-orbit interaction is due to the **B** field parallel to **L**, which is imposed by Lorentz transformation of the Coulomb field of the nuclear charge.

the Zeeman energy E_{mag} of a magnetic moment μ is

$$E_{mag} = -\mu \cdot \mathbf{B} \tag{6.79}$$

In the description of the atoms, no magnetic fields emerge, as long as the velocities of the electrons are non-relativistic. In this case, the potential energy is calculated from an integral of the Coulomb potentials of the individual charges in an atom. If the velocities of the electrons approach the speed of light the **B** field due to the Lorentz transformation of the Coulomb field of the nucleus increases and perturbs the purely electrostatic picture. For electrons with no angular momentum (s electrons), the average magnetic field is zero which cancels the effect. The same applies to closed atomic shells because the effect on the up electrons is canceled by that on the down electrons. For open shells with angular momentum, we get a **B** field, perpendicular to the motion of the electron and the radius, or parallel to the angular momentum. The B-field in the center of a loop with radius r on which a charge Ze is rotating in an orbit is

$$B_z = \frac{\mu_0 Ze}{4\pi r^2} v\,, \tag{6.80}$$

which is the B-field in the rest frame of the electron. With the electric field at the site of the electron E_r in the rest frame of the nucleus:

$$E_r' = \frac{Ze}{4\pi\epsilon_0 r^2}\,, \tag{6.81}$$

and with $\epsilon_o \mu_o = 1/c^2$ we get:

$$B_z = -\frac{\beta}{c} E_r'\,. \tag{6.82}$$

This B-field is proportional to the angular momentum L_z, and with Eq. (6.79) we get for the spin-orbit interaction:

$$E_{SO} \propto \ell \cdot \sigma\,, \tag{6.83}$$

where ℓ and σ are the angular momentum and the spin quantum numbers, which are 1 and 1/2 for the case of a $2p$ electron wave function. In Figure 6.15 the spin-orbit interaction due to the coupling of the spin magnetic moment to the magnetic field of the moving charge is visualized.

7 Space travel

Exploration of unknown regions is a major motivation for human activity and interest, as is demonstrated by the popularity of science fiction. Space is perhaps the last frontier, and the first steps to its exploration with automated probes and manned spaceflights have been made. This Chapter is for friends of science fiction and those who would like to experience the opportunities and consequences of special relativity with their own body. We refrain from super-luminal space travel, but work out the possibilities and limitations of space travel within the framework of special relativity. We probe the limits of interstellar flight, both from the problem of spacecraft propulsion and that of the dangers in so-called *empty space* at relativistic velocities.

7.1 Can we cross the universe

The distances to stars and galaxies are so large that thousands to millions of years are necessary to reach them with present-day technology. Only with a constantly accelerated flight, the crew on board a spacecraft can reach distant stars and galaxies within a lifetime of less than one hundred years. Although this is possible in principle, it comes at the prize that the world outside of the ship ages by millions of years.

7.1.1 Requirements

Interstellar or intergalactic space travel involves special relativity and the duration of the journeys depends on whether the time is measured on Earth or in the spacecraft. Interplanetary travel is within reach of presently conceivable technology and seems to be feasible considering the requirements of energy and duration of the voyage. The longer distances in interstellar traveling require more energy and time. The absolute limit of achievable velocities makes a trip even to the stellar neighbourhood of Earth very long. This is certainly true for the people staying on Earth and waiting for the return of the astronauts. The proper time for the astronauts is dilated if compared with the time on Earth. When traveling at relativistic velocities they would return to an Earth where all their friends are gone long ago.

For interstellar travel, highly relativistic velocities are necessary, and thus, steadily accelerated flight is required in order to reach these velocities. The most comfortable for the astronauts would be a uniform proper acceleration of $c\,\alpha$=10 m/s^2, corresponding to Earth's gravity (see Section 3.1.2). In units appropriate for space flight, this is about $c\,\alpha$=1 ly/a^2 (light-year/year^2), measured in proper distance and proper time on the spacecraft. After one proper year of acceleration, the spacecraft reaches the rapidity $\rho = 1$. If the trip shall be related to the space that is crossed, the corresponding proper velocity ω,

https://doi.org/10.1515/9783111503592-008

which measures the distance traveled in the rest frame of Earth per given proper time in the spacecraft applies, see Figure 3.2,

$$\omega = \sinh(\rho) = \frac{1}{2}\left(e^{\rho} + e^{-\rho}\right) . \tag{7.1}$$

Thus, with constant proper acceleration and for large rapidities, the proper velocity increases exponentially with rapidity and proper time τ,

$$\omega \approx e^{\rho} = e^{a\tau} . \tag{7.2}$$

This result is encouraging: in principle, it is possible to travel cosmic distances within the lifetime of human beings provided that a reasonably large proper acceleration during the entire journey is maintained.

The realization of such a cosmic travel plan requires the solution of some serious problems, of which most are based on relativistic effects. The first problem concerns the necessary specification for the spacecraft that is able to accelerate over a long time in outer space. There are two different scenarios: the first is a rocket that carries all necessary fuel from the start, and the second is the ramjet (Bussard, 1960) that collects interstellar matter that is used as a fuel supply. Possible spacecrafts are discussed in Section 7.2.

More problems are caused by the existing radiation in space and the interstellar or intergalactic matter. Although the spacecraft travels through almost empty space, problems are caused by the extreme Doppler shift of the electromagnetic radiation at highly relativistic velocities, and by residual matter like hydrogen, helium, and dust particles. These obstacles are discussed in Section 7.4. Before this, we will discuss the implications of uniformly accelerated flight in different scenarios.

7.1.2 Uniformly accelerated flight

Consider the kinetics (see Section 3.1) for a flight to α Centauri or the Andromeda galaxy with constant proper acceleration a. If the velocity of the astronaut at the start is $\beta_0 = \omega_0 = \rho_0 = 0$ at time $t_0 = \tau_0 = 0$. With Eq. (3.3), the distance traveled in the frame of reference at the beginning of the trip is

$$s = \int_0^s ds' = c \int_0^\tau \omega(\tau')\, d\tau' = \frac{c}{a}\left[\cosh(a\tau) - 1\right] . \tag{7.3}$$

Conversely, the proper time needed to travel the distance s in the reference system is

$$\tau = \frac{1}{a} \operatorname{arccosh}\left(1 + \frac{as}{c}\right) . \tag{7.4}$$

Integration of Eq. (3.17) yields the needed coordinate time to reach the proper velocity ω,

$$t = \frac{1}{a} \int_0^\omega d\omega' = \frac{\omega}{a} = \frac{1}{a} \sinh(a\tau) . \tag{7.5}$$

Eqs. (7.4) and (7.5) also give the distance traveled by constant proper acceleration as a function of coordinate time,

$$s = \frac{c}{\alpha}\left(\sqrt{1+\alpha^2 t^2} - 1\right), \tag{7.6}$$

and reversely, solving Eq. (7.6) yields the time t as a function of the traveled distance s,

$$t = \frac{1}{c}\sqrt{\frac{s\,(\alpha s + 2c)}{\alpha}}\,. \tag{7.7}$$

Eqs. (7.3) and (7.5) constitute a parameter equation (parameter τ) for the world line of a uniformly accelerated astronaut. Elimination of the proper velocity ω yields the coordinate equation,

$$\left(s + \frac{c}{\alpha}\right)^2 - (ct)^2 = \left(\frac{c}{\alpha}\right)^2, \tag{7.8}$$

which is the equation of a hyperbola in a Minkowski diagram. The asymptotes of the hyperbola have inclinations ± 1 and are parallel to the light cone. This has an interesting consequence: an astronaut starting at $s = 0$ moving with constant proper acceleration α in the direction of the positive s-axis outruns a photon starting simultaneously at $s \leq -c/\alpha$ in the same direction (Misner et al., 1973). $s = s_{\mathrm{h}} = -c/\alpha$ signifies a communication horizon whose consequences are also discussed in the Section on communication (7.3.3).

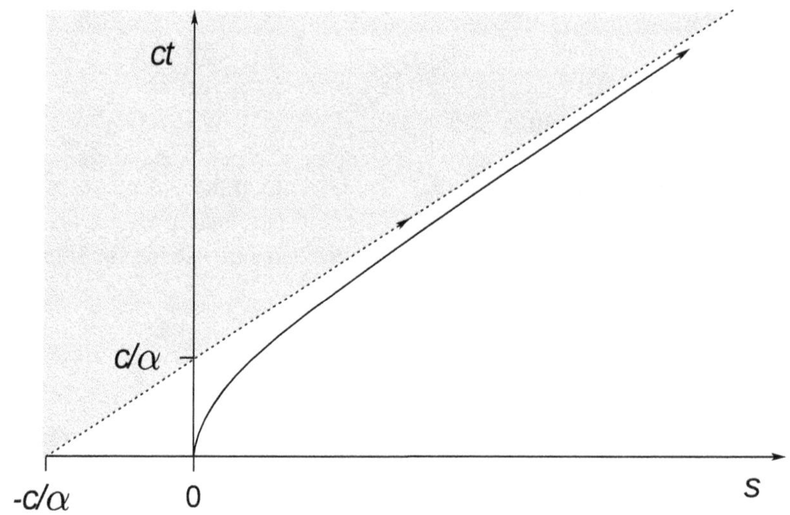

Fig. 7.1: World line of a spacecraft flying with constant proper acceleration α (solid line) and of a photon (dashed line), both starting at the same time on the s-axis. The shaded area lies behind the communication horizon, from where no information can reach the spacecraft. (Adopted from (Greber and Blatter, 2006)).

Horizons divide space such that no information from one point in one part can reach the other part. Such a horizon exists for an astronaut who travels with constant proper acceleration. No photons or any other signal from the grey region in Figure 7.1 can reach the astronaut at any time, as long as the acceleration lasts. Thus, the position

$$s_{\mathrm{h}} = c\,t - \frac{c}{\alpha} \tag{7.9}$$

marks the moving horizon behind the traveler, and no signal transmitted from the starting point after time c/α, can reach him anymore. The space travel scenario $c\,\alpha = 1$ g implies that after one year the spacecraft may obtain no more information from the starting point. On the other hand, the signals sent out by the astronaut can reach any point in space, though the signals may be extremely Doppler-shifted.

To reach interstellar distances within a reasonable time, high velocities, and thus, continuous acceleration during travel time is required. A flight to a star must include two stages: acceleration to high velocities and breaking down the velocity to the velocity of the star. At a constant proper acceleration of 9.8 m/s^2, corresponding to about 1 light-year/year^2, a flight to α Centauri in 4.3 light-years distance will last about 3.63 proper years. In the middle between Earth and α Centauri, when the thrust is reverted, the rapidity, Eq. (3.9), of the spacecraft would be $\rho = 1.4$, the proper velocity $\omega = 1.9$ and the coordinate velocity relative to Earth $\beta = 0.88$. A flight to α Centauri and back to Earth would last at least 7.26 proper years, whereas on Earth 14.5 years would pass between take off and return of the spacecraft. The corresponding time to travel to the Andromeda galaxy at a distance of $2.5 \cdot 10^6$ light-years and back is 58 proper years. On Earth, about 8 Million years elapse during the corresponding trip to the Andromeda galaxy and back. At this constant proper acceleration, the covered distances become large at a large rate. In 20 proper years, the astronaut covers a distance of 250 Million light-years and 36 Billion light-years after 25 proper years.

Therefore, it is, in principle, possible to travel to the most distant galaxies within a human lifetime, as is outlined quite dramatically in the science fiction novel "Tau Zero" by Poul Anderson (1970), if such a spacecraft would be available: During the first part of the journey, from the start on Earth to the midpoint between Earth and Andromeda, the spacecraft is accelerating with $c\,\alpha = 1$ g, and its rapidity at the midpoint is $\rho = 14.5$, $\omega = 991380$ and $\beta = 0.9999999999995 = (1 - 5 \cdot 10^{-13})$. Midway, the astronauts may want to look out for their goal, the Andromeda galaxy, or look back to the origin of their journey, the Milky Way galaxy. An observer at this midpoint at rest with respect to the galaxies would see the disks of the galaxies in cones with opening angles of $\theta_A = 4.9°$ for Andromeda and $\theta_M = 2°$ for the Milky Way galaxy. Due to aberration, for the astronauts, moving with $\rho = 14.5$, the corresponding aspect angles are $\theta'_A = 0.000006°$ and $\theta'_M = 179.8°$ for Andromeda and the Milky Way, respectively. The Andromeda galaxy shrinks to a tiny dot during the accelerated approach, whereas the Milky Way spans almost all of the sky. While the light of the Milky Way is Doppler shifted by a factor of 10^6 to extremely faint microwave radiation, the Andromeda galaxy appears as an extremely tiny but intensive γ-ray source ahead of the spacecraft. We have thus the seeming paradox situation that an object ahead of the flight appears to become smaller and smaller during an accelerated approach and an object behind seems to grow larger and larger during the flight (Beig and Heinzle, 2008). Aberration by far dominates the aspect angle over the effect of distance.

7.2 Spacecrafts

Space traveling requires *spacecrafts* that can accelerate. For this it makes use of momentum conservation and exhaust or harvest of momentum. In empty space acceleration bases on the concept of *rockets* which carry the propellant for accelerating it into exhaust from the beginning of the journey. If space is not completely empty *spaceships* can be operated. They harvest the energy for propulsion at their actual location in space and time. A ship may be propelled by solar wind or solar light. These sources of energy and momentum rely on nearby stars and are limiting the reach of deep space. If the fuel is interstellar matter, it can be collected and used as a fuel for propulsion. This form of spaceship is discussed in Section 7.2.4 as *ramjets*.

7.2.1 Rocket equations

A rocket is carrying both, the payload and the propellant from the beginning of the journey. To accelerate in a desired direction, some matter must be ejected in the opposite direction, while the conservation of momentum applies.

The equation for the acceleration of a rocket is derived using the view of the spacecraft, i.e. the quantities observed in the spacecraft, compare Section 7.2.3. The measured acceleration of the rocket is the proper acceleration α (Section 3.1.2), and the rapidity ρ is obtained by integrating α over the proper time τ. The exhaust velocity corresponds to the coordinate velocity relative to an inertial coordinate-system momentarily co-moving with the rocket.

The total mass $M_0 = m_0 + m$ of the rocket at the beginning of the journey consists of the mass m_0 of the payload and the mass m of the propellant. At a constant exhaust rate of propellant, the mass of the rocket at proper time τ is

$$M(\tau) = m_0 + m - \mu \dot{N} \tau \,, \tag{7.10}$$

where $\dot{N}$ is the ejection rate of propellant particles with mass μ, and

$$T_{\text{tot}} = \frac{m}{\mu \dot{N}} \tag{7.11}$$

is the total burning time of the rocket.

First, a non-relativistic exhaust velocity β_s is considered, thus the equation of motion in the system of the rocket can be written in classical form,

$$Ma + \frac{p}{c}\dot{N} = 0\,, \tag{7.12}$$

where $p = \mu\, c\, \beta_s$ is the momentum of an exhaust particle. Integration over the entire burning time T_{tot} of the proper acceleration $a(\tau)$ from Eq. (7.12) yields a maximum change in rapidity

$$\Delta\rho = -\frac{p}{c}\int_0^{T_{\text{tot}}} \frac{\dot{N}}{m_0 + m - \mu\,\dot{N}\,\tau}\, d\tau = -\frac{p}{\mu\, c}\ln\xi = -\beta_s \ln\xi \tag{7.13}$$

where $\xi = m_0/(m_0 + m) \leq 1$ is the *payload ratio*. The reachable final rapidity change does not depend on $\dot{N}$, but on the exhaust velocity and the payload ratio. In the limit for $\xi \to 0$, $\Delta\rho$ diverges, and in principle, any desired final rapidity could be reached with a conventional rocket, even if the exhaust velocity is non-relativistic, $|\beta_s| \ll 1$. Still, a calculation for a payload of 1000 kg and an exhaust velocity of 300 km s^{-1} shows that much more than the mass contained in the visible universe as propellant and much more time than the age of the universe would be needed in order to reach 1% of the speed of light.

Conventional rockets are high-thrust systems but use low-specific momentum propulsion systems based on chemical reactions. The exhaust velocities of the gas are limited by aerodynamic properties, mainly by the speed of sound waves in the gas. Such systems are required for spacecrafts to lift heavy loads within a short time into orbit or to escape velocities. Once in space, low thrust high specific momentum systems working for long periods of time may accelerate the spacecraft to much higher velocities over some period of time and finally, save a lot of time for interplanetary missions. To operate a propulsion system efficiently, high exhaust velocities are required for long periods of time.

The situation does not look much better for relativistic exhaust velocities $\beta_s \to 1$. For this case, μ in Eq. (7.10) must be replaced by the mass equivalent μ_E of the total energy E_{tot},

$$E_{\text{tot}} = \mu_E\, c^2 = E_{\text{kin}} + \mu\, c^2\,, \tag{7.14}$$

and thus

$$\beta_s \equiv \frac{p}{\mu_E\, c} = \frac{pc}{E_{\text{tot}}}\,. \tag{7.15}$$

With Eq. (3.43), we obtain for the relativistic exhaust velocity of massive particles

$$\beta_s = \frac{\sqrt{E_{\text{kin}}^2 + 2E_{\text{kin}}\,\mu\, c^2}}{E_{\text{kin}}^2 + 2\mu\, c^2} < 1. \tag{7.16}$$

At best the exhaust velocity can reach $\beta_s = 1$ if $\mu = 0$, i.e. for massless photons. A relativistic rocket emitting photons could in principle gain the largest $\Delta\rho$ for a given payload ratio ξ. Since nuclear fusion can only convert less than 1% of the mass into photons, ξ can only be little smaller than one and therefore $\Delta\rho$ would be very small for a photon propulsed system.

7.2.2 Electrostatic ion propulsion systems

Electrostatic ion thrusters use the electric force on charged particles to accelerate these particles in an electric field. The ionized gas, often the heavy noble gas xenon, is ionized and then injected into a chamber where the positive ions are attracted by a gridded cathode (negative pole), through which the ions are ejected. The ion beam must be neutralized with electrons to prevent the spacecraft from being loaded with negative charge (Figure 7.2).

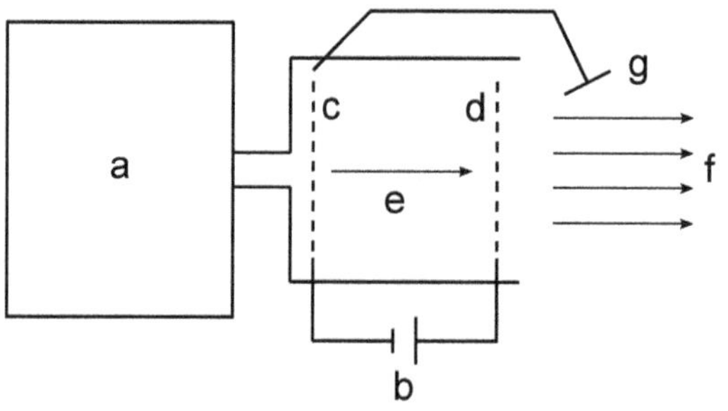

Fig. 7.2: Schematic of an ion propulsion system: from a reservoir the gas (e.g. xenon) is ionized and injected into an electric field between two electrodes c and d connected by a high voltage power source b. The accelerated ions are ejected f and neutralized by an additional electrode g to prevent an electric charging of the whole device.

The first ion propulsion system was tested successfully on the Deep Space 1 (DS1) mission of the National Aeronautics and Space Administration (NASA):

> Deep Space 1 (DS1) launched from Cape Canaveral on October 24, 1998. During a highly successful primary mission, it tested 12 advanced, high-risk technologies in space. In an extremely successful extended mission, it encountered Comet Borrelly and returned the best images and other scientific data ever from a comet. During its fully successful hyperextended mission, it conducted further technology tests (Brophy et al 2000). The spacecraft was retired on December 18, 2001 .

The electrostatic thruster NSTAR on Deep Space 1 had a length of 30 cm, a weight 8 kg and a variable thrust of 20 to 92 mN. The total mass of the spacecraft at launch was

486.3 kg, composed of 373.4 kg dry spacecraft, 31.1 kg of hydrazine, and 81.5 kg of xenon propellant, which provided about 20 months of continuous thrusting. The ion thruster on Deep Space 1 changed the spacecraft's speed by 4'500 m/s.

7.2.3 Electromagnetic thrusters

The first part of the abstract in Bering et al. (2007) formulates the three main principles applied for the electromagnetic propulsion system:

> The Variable Specific Impulse Magnetoplasma Rocket (VASIMR) is a high-power magnetoplasma rocket, capable of Isp/thrust modulation at constant power. The plasma is produced by a helicon discharge. The bulk of the energy is added by ion cyclotron resonance heating (ICRH.) Axial momentum is obtained by adiabatic expansion of the plasma in a magnetic nozzle. ...

The method of heating plasma used in VASIMR was originally developed as a result of research on nuclear fusion reactors. VASIMR is intended to bridge the gap between high-thrust, low-specific momentum propulsion systems and low-thrust, high-specific momentum systems as it is capable of functioning in either mode.

The VASIMR thruster consists of 4 parts (Figure 7.3). The propellant, argon gas, is first injected into a chamber where it is ionized with electromagnetic wave heating using the so-called helicon (low frequency) electromagnetic antenna system. This *cold plasma* at about 6000 K temperature then flows into a magnetic bottle where its temperature is further increased to more than a million degrees by resonant electromagnetic wave heating. The ions with the highest velocity relative to the magnetic field lines eventually escape and are further accelerated by the fact that the magnetic field lines diverge outside the magnetic bottle and the helical trajectories of the ions elongate and thus accelerate in the exhaust direction. The final velocity of the argon ions is about 50'000 m/s.

The velocity of the plasma escaping the magnetic bottle corresponds to the thermal velocity of the particles at the corresponding temperature. Since the electrons are much lighter than the ions, their thermal velocity is much higher. A separation of the negatively and positively charged particles produces an electric field that tends to accelerate the slower ions and decelerate the faster electrons until they move at the same velocity as a so-called quasi-neutral plasma. This so-called *ambipolar diffusion* eventually results in a neutral exhaust gas.

The VASIMR thruster VX-200 is planned to operate at a radio frequency power of 200 kW and deliver a thrust of 5.9 N.

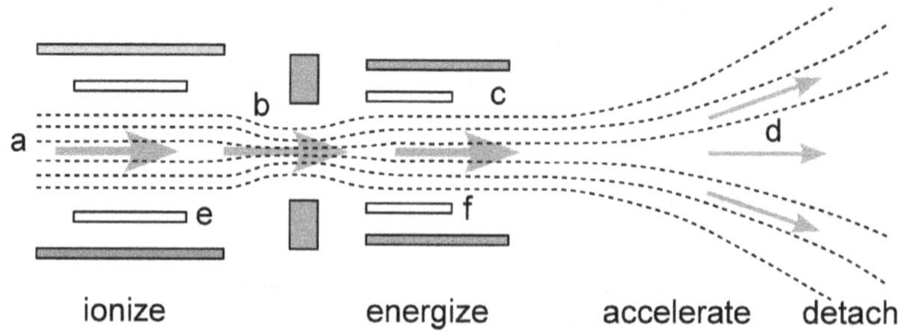

Fig. 7.3: Schematic of an electromagnetic ion thruster: the injected gas (a) is heated and ionized by electromagnetic waves (antenna e). The energetic particles pass the bottleneck in magnet b and are further energized by electromagnetic waves (antenna f) and after leaving the device are further accelerated along the propulsion direction by the diverging magnetic field lines behind the device.

7.2.4 Ramjets

The results for rocket propulsion demonstrate the infeasibility of propulsion with on-board fuel for interstellar spaceflight. Bussard (1960) forwarded the idea to use interstellar matter for propulsion in a ramjet-type spacecraft. The Bussard ramjet (Figure 7.4) collects interstellar matter through a intake funnel and uses the hydrogen for fusion and propulsion. The idea gained much interest in space technology and science fiction and produced several works discussing the principle and technical difficulties and the possible limits of Bussard ramjet spaceflights. In the following, the basic physics of the Bussard ramjet is introduced and discussed (Blatter and Greber, 2017).

We describe the flight of a Bussard ramjet from the position of an astronaut in the spacecraft. The collection of mass in interstellar or intergalactic space depends on the density ϱ of the gas, of the collecting area A of the ramjet and the distance Δz, which the ramjet flies with the proper velocity ω in the proper time interval $\Delta\tau$,

$$\Delta m = \varrho A\,\Delta z = \varrho A\,c\,\omega\,\Delta\tau\,. \tag{7.17}$$

Ramjets stopping the incoming matter

The momentum balance of a ramjet, which is stopping the collected mass and then burns it in a fusion reactor is

$$p_2 - p_1 = \Delta p = p_{\mathrm{m}} - p_{\mathrm{f}}\,, \tag{7.18}$$

where p_1 and p_2 are the momenta of the ramjet before and after a given time interval $\Delta\tau$. p_{m} is the exhaust momentum transferred to the ramjet, and p_{f} the momentum due to the friction or stopping of Δm. We get

$$p_1 = M\,\omega_1\,c \quad , \quad p_2 = M\,\omega_2\,c \tag{7.19}$$

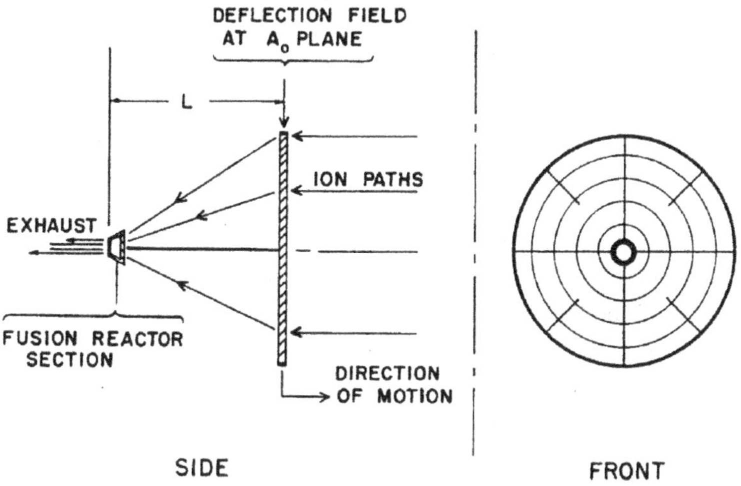

Fig. 7.4: Schematic drawing of a ramjet as anticipated by Bussard (1960).

where M is the mass of the ramjet, and

$$p_\mathrm{m} = (1-\epsilon)\,\Delta m\,\omega_\mathrm{m}\,c \quad , \quad p_\mathrm{f} = -\Delta m\,\omega_1\,c\,, \tag{7.20}$$

with ω_m the proper velocity of the exhausted gas and ϵ the mass defect due to the gain of fusion energy. Finally, the momentum balance in Eq. (7.18) becomes

$$M\,(\omega_2 - \omega_1) = (1-\epsilon)\,\Delta m\,\omega_\mathrm{m} - \Delta m\,\omega_1\,. \tag{7.21}$$

For the terminal velocity ω_∞, i.e. for $\Delta p = 0$ we get,

$$\omega_1 = \omega_2 = \omega_\infty = \omega_\mathrm{m}\,(1-\epsilon)\,, \tag{7.22}$$

or with Eq. (3.6)

$$\gamma_\mathrm{m} = \frac{\sqrt{(1-\epsilon)^2 + \omega_\infty^2}}{1-\epsilon}\,. \tag{7.23}$$

The apparent singularity for $\epsilon \to 1$ disappears for the determination of ω_∞, when we consider the kinetic energy balance for the solution of the kinematic problem

$$E_2 - E_1 = \Delta E = E_\mathrm{fus} - E_\mathrm{exh}\,, \tag{7.24}$$

where E_1 and E_2 are the kinetic energies of the ramjet before and after the time interval $\Delta\tau$. E_fus is the fusion energy that can be transformed into kinetic energy of the exhausted mass,

$$E_\mathrm{fus} = \epsilon\,\Delta m\,c^2\,, \tag{7.25}$$

and E_exh is the kinetic energy in the exhaust,

$$E_\mathrm{exh} = (\gamma_\mathrm{m} - 1)\,(1-\epsilon)\,\Delta m\,c^2\,. \tag{7.26}$$

If the ramjet reaches its terminal velocity it does not change its velocity and $\Delta E = 0$. From $E_{\rm fus} = E_{\rm exh}$ we get a second expression for $\gamma_{\rm m}$

$$\gamma_{\rm m} = \frac{1}{1-\epsilon}\,. \tag{7.27}$$

With Eq. (7.23) the singularities for $\epsilon \to 1$ cancel and we finally get the terminal velocity

$$\omega_\infty = \sqrt{\epsilon\,(2-\epsilon)}\,, \tag{7.28}$$

which can not exceed $\omega_\infty = 1$. With a mass defect $\epsilon = 0.007$, the final velocity of the ramjet is $\omega_\infty = 0.118$ or about 12% of the speed of light. For this reason, it is necessary to avoid the friction caused by the collection and the stopping of the mass of the interstellar fuel if velocities $\omega > 1$ shall be obtained.

The ideal Bussard ramjet

The *ideal Bussard ramjet* is a spacecraft, in which the entire collected mass is funneled through the jet-engine without stopping it, though fuse it and accelerate the fusion products in one direction.

The momentum balance in Eq. (7.21) for the ramjet that stops the incoming mass without the stopping term reduces to

$$M\,(\omega_2 - \omega_1) = M\,c\,\Delta\omega = (1-\epsilon)\,\Delta m\,\omega_{\rm m}\,. \tag{7.29}$$

From Eq. (7.18) we obtain the proper velocity of the exhausted gas,

$$\omega_{\rm m} = \frac{M\,\Delta\omega}{(1-\epsilon)\,\Delta m} \tag{7.30}$$

and with this, the relativistic factor,

$$\gamma_{\rm m} = \frac{\sqrt{(1-\epsilon)^2\,\Delta m^2 + M^2\,\Delta\omega^2}}{(1-\epsilon)\,A\varrho c\omega\,\Delta\tau}\,, \tag{7.31}$$

which we will use in the energy balance equation below to eliminate $\gamma_{\rm m}$.

The energy balance equation consists of the energy gain of the ramjet, the energy content of the collected mass, which can be gained by fusion, and the kinetic energy of the exhausted mass,

$$M(\gamma_2 - \gamma_1) - \epsilon\,\Delta m + (1-\epsilon)(\gamma_{\rm m} - 1)\,\Delta m = 0\,. \tag{7.32}$$

The difference $\Delta\gamma = \gamma_2 - \gamma_1$ is related to the proper acceleration $\alpha = \Delta\omega/\Delta\tau$ of the ramjet by

$$\Delta\gamma = \frac{\omega\,\Delta\omega}{\gamma} = \omega\,\alpha\,\Delta\tau\,, \tag{7.33}$$

With this, the energy balance equation becomes

$$M\,\alpha + \gamma_{\rm m}\,(1-\epsilon)\,\varrho A\,c - \varrho A\,c = 0\,. \tag{7.34}$$

and with Eq. (7.31), we can eliminate $\gamma_{\rm m}$ in Eq. (7.34),

$$\omega\,\Delta\tau\,(M\,\alpha - \varrho A\,c) = -\sqrt{(1-\epsilon)^2\,\Delta m^2 + M^2\,\Delta\omega^2}\,. \tag{7.35}$$

By taking the square, we obtain

$$M^2\,\alpha^2 - 2M\,\alpha\varrho A\,c + (\varrho A\,c)^2 = \frac{(1-\epsilon)^2\,\Delta m^2}{\omega^2\,\Delta\tau^2} + \frac{M^2\,\Delta\omega^2}{\omega^2\,\Delta\tau^2}\,. \tag{7.36}$$

The two terms on the right side are

$$\frac{(1-\epsilon)^2\,\Delta m^2}{\omega^2\,\Delta\tau^2} = (1-\epsilon)^2\,(\varrho A\,c)^2 \tag{7.37}$$

and

$$\frac{M^2\,\Delta\omega^2}{\omega^2\,\Delta\tau^2} = \frac{\gamma^2}{\omega^2}\,M^2\,\alpha^2 = \frac{1+\omega^2}{\omega^2}\,M^2\,\alpha^2\,. \tag{7.38}$$

If we expand the $(1-\epsilon)^2$ and omit the small ϵ^2 term, we obtain a quadratic equation for the energy of the ramjet,

$$M^2\,\alpha^2 - 2M\,\alpha\,\omega^2\varrho A\,c + 2\,\epsilon\,(\varrho A\,c\,\omega)^2 = 0\,. \tag{7.39}$$

The solution of Eq. (7.39) is straightforward,

$$\alpha = \frac{\varrho\,A\,c}{M}(\omega^2 - \omega\sqrt{\omega^2 - 2\,\epsilon})\,, \tag{7.40}$$

which is the achievable acceleration for the ramjet at a given proper velocity. With the approximation

$$\sqrt{\omega^2 - 2\,\epsilon} = \omega\sqrt{1 - \frac{2\,\epsilon}{\omega^2}} \approx \omega\left(1 - \frac{\epsilon}{\omega^2}\right)\,, \tag{7.41}$$

the limit acceleration α_∞ for large proper velocities is given by

$$\alpha_\infty = \epsilon\,\frac{A\varrho c}{M}\,. \tag{7.42}$$

From this it is seen that the ideal Bussard ramjet without stopping the mass for propulsion and without friction may endlessly accelerate.

This result suggests the possibility of a journey in space described in the science fiction novel "Tau Zero" by Poul Anderson (1970), in which a Bussard ramjet is doomed to (almost) endlessly accelerate. Later, serious limitations and even the infeasibility of Bussard ramjets were discussed, including energy loss by radiation (Semay and Silvestre-Brac, 2007, 2008) or the structural strength of the construction of a large device collecting the interstellar matter. In this work, we do not discuss the feasibility of the Bussard ramjet but the kinematic implications of a trip with such a hypothetical vehicle.
Eq. (7.42) allows to estimate the size of the ramjet for the given requirements. For a trip across the universe, an acceleration of $c\,\alpha_\infty = 10\ \mathrm{ms}^{-2}$, corresponding to Earth gravitational acceleration g, is comfortable. Let us assume the weight of the ramjet corresponds

to the weight a carrier ship, $M = 10^8$ kg. The area of the collecting shield or funnel then must be

$$A = \frac{a_\infty M}{\epsilon \varrho c}, \tag{7.43}$$

where we note that the density ϱ of the interstellar or intergalactic gas varies many orders of magnitude. In interstellar clouds, there may be up to 1000 hydrogen atoms per cubic centimeter (Yeghikyan and Fahr, 2004), corresponding to a density $\varrho = 4 \cdot 10^{-19}$ kg m^3, in the local interstellar medium there are about 5 atoms per cubic centimeter (Cox and Reynolds, 1987), corresponding to a density of $\varrho \approx 2 \cdot 10^{-21}$ kg m^3. Considering the average baryon density of the universe, the critical density for a flat universe is between 1 and $3 \cdot 10^{-26}$ kg m^3 (Spergel et al., 2003). For the local interstellar medium, according to Eq. (7.43), the area of the collecting funnel must equal to about the area of the cross-section of the Earth. Although a single layer of carbon as it is graphene is the lightest material known today to be used for such a funnel a single layer of graphene of Earth-size has already a mass of 10^8 kg.

7.3 Challenges

Also for a cosmic voyage, orientation, navigation and communication are fundamental in order to reach a desired destination. *Orientation* means to firstly determine the position in space. Furthermore, the direction of motion must be determined. Based on the information available, such as a four-dimensional, eventually dynamical, map of the stars or galaxies and the cosmic background radiation, a hierarchy of different problems can be formulated. *Navigation* is the manoeuvring of a spacecraft as it approaches its destination, including observing and avoiding obstacles or low-density space that inhibits propulsion. *Communication* also becomes a challenge, mainly because of the Doppler shift, where accelerated space travel may even create horizons beyond which no information can be sent.

7.3.1 Orientation

Orientation during a space flight includes two major tasks, firstly, the determination of the position at a given time, and secondly, the determination of the velocity vector. Both require a reference frame, desirably fixed in space if this can be a well-defined quality. Since about every object in space is in motion, absolute references are difficult to define. At least the angular motion of objects becomes slower with increasing distance. Thus, a frame of reference of such “fix stars” is stable for a “long” time. However, during a spaceflight with hyperrelativistic velocities, the term “long” may become “short” in view of the extreme time dilation.

In the following we assume that a stable frame of reference is defined and can be recalled at any time for orientation purposes. One possibility to define a frame of reference is a spherical coordinate system centered at the spacecraft. The colatitude θ is measured with respect to a defined pole on the celestial sphere and the longitude φ is measured with respect to a reference longitudinal circle.

To determine a position in space with respect to a field of stars or galaxies, the distances of the stars, and galaxies must be known. Once a sufficient number of stars or galaxies are mapped on the celestial sphere, including their distances, a comparison with a three dimensional map of stars and galaxies of the surrounding space would in principle be sufficient to determine the actual position and the direction of motion with respect to the background rest frame. The cosmic background radiation is isotropic except for small-scale patterns. Any motion with respect to this frame is reflected in a dipole anisotropy, which allows us to determine the speed and the direction of this motion.

Since stars move with respect to their galaxy and galaxies move with respect to the "cosmic" frame, the three-dimensional map of the stars and galaxies must be dated in a four-dimensional map. This may not be important if the astronaut moves with low relativistic velocities. At extreme relativistic velocities, even a dated map that assumes no interaction may remain useful for the entire journey. Time dilation can be so large that a substantial acceleration of the galaxies take place within the proper lifetime of the astronaut.

Cosmic microwave background radiation

Viewed from Earth, the cosmic microwave background radiation appears to be an ideal black body radiation with an average temperature $T_{\mathrm{av}} = 2.728 \pm 0.004$ K. The angular distribution shows a pronounced dipole characteristics with a temperature difference of 3.372 ± 0.014 mK between the poles and the mean temperature (Fixsen et al., 1996). Since the radiation temperature is transformed like a frequency we get with Eq. (2.24)

$$T'_{\mathrm{antapex}} = T_{\mathrm{av}} \sqrt{\frac{1-\beta}{1+\beta}}, \quad T'_{\mathrm{apex}} = T_{\mathrm{av}} \sqrt{\frac{1+\beta}{1-\beta}}, \tag{7.44}$$

and obtain a velocity relative to the background $\beta = 6.2 \cdot 10^{-4}$ ($v = 185.2 \pm 0.8\ \mathrm{km\,s^{-1}}$), while the direction of the apex lies in the constellation of Leo.

It seems that there exists a local frame where the background appears isotropic and which can be considered to be the local cosmic frame. A map of the cosmic background thus can serve as a means to determine the velocity vector of a spacecraft with respect to this frame. A complete map of the background temperature is laborious to obtain, thus, we may be interested in the minimal number of measurements in different directions required to determine the velocity vector. To determine the orientation of a dipole as described in Eqs. (7.44), where the orientation of the dipole axis is unknown, in an otherwise isotropic universe we need the background temperature measurements in four different directions. This will give the average temperature T_{av}, the velocity relative to the background β and the two spherical angles θ and φ for the dipole orientation.

The temperature or the frequency of the background radiation, transforms as

$$\frac{1}{T'} = \frac{1}{T_{av}} \gamma (1 - \beta \cos \theta') . \tag{7.45}$$

see Eqs. (6.26) and (6.29), where T' is the observed temperature of the background radiation at a given point of the celestial sphere and T_{av} is the isotropic temperature and defines the rest frame of the background radiation.

One possibility to use the angular distribution of the cosmic background temperature for a determination of one's direction of motion is given by the fact that points with the same temperature T' lie on a circle of constant θ'. With this fact, the problem can be shifted from a mathematical problem to a problem of the observational method to find points with equal radiation temperature. Once three different points are identified, the circle through the points can be determined by means of elementary geometry and vector algebra. The spherical center of this circle is either the apex or the antapex of the motion, but a fourth temperature different from the other three is required to obtain complete information on the dipole field. In particular, the radiation temperatures of the now-known apex and antapex give the magnitude of the velocity in the most direct way.

In the following, the mathematical details of this determination of the velocity vector is given. Firstly, a spherical coordinate system with a pole and an equator must be defined. Then, the three points are given by their respective polar angles θ_i and longitudes φ_i on the celestial sphere, $P_i(\theta_i, \varphi_i)$, $i = 1, 2, 3$ (see Section 2.1). We introduce Cartesian coordinates with the origin in the center of the sphere, the z-axis pointing to the pole with $\theta = 0$ and the x-axis pointing in the direction $\varphi = 0$. Assuming that the radius of the celestial sphere is unity, the Cartesian coordinates of the three points P_i are

$$x_i = \sin \theta_i \cos \varphi_i , \quad y_i = \sin \theta_i \sin \varphi_i , \quad z_i = \cos \theta_i . \tag{7.46}$$

With the three points, two independent vectors can be defined, e.g. vector $\mathbf{a}$ points from P_1 to P_2 and vector $\mathbf{b}$ points from P_1 to P_3. The normal vector $\mathbf{n}$ of the plane defined by the three points is given by the cross product $\mathbf{n} = \mathbf{a} \times \mathbf{b}$. The parameter representation of the straight line parallel to $\mathbf{n}$ through the origin O(0,0,0) of the Cartesian coordinate system is given by

$$\mathbf{r} = t\,\mathbf{n} , \tag{7.47}$$

where $\mathbf{r}$ is the variable position vector of points on the straight line and t is a parameter. The sections of the line with the sphere are the poles and either the apex or the antapex, which are the endpoint of the vectors $\mathbf{r}_a$ and -$\mathbf{r}_a$

$$\mathbf{r}_a = \frac{\mathbf{n}}{|\mathbf{n}|} = \begin{pmatrix} x_a \\ y_a \\ z_a \end{pmatrix} . \tag{7.48}$$

Finally, the three Cartesian coordinates of the poles can be transformed back to spherical coordinates in Eq. (7.46)

$$\cos\theta_a = z_a\,, \quad \sin\varphi_a = \frac{y_a}{\cos\theta_a}\,, \quad \cos\varphi_a = \frac{x_a}{\sin\theta_a}\,. \tag{7.49}$$

Both equations for the longitude of the apex or antapex, φ_a, are necessary to uniquely determine the quadrant of the angle. The apex is discriminated from the antapex via their different temperatures. Once the apex is given, the absolute value of the velocity can be computed with Eq. (7.44).

It seems to be easier to measure the temperature of a chosen point in the sky than to find a point with a given temperature. The above considerations suggest that 4 points of different temperatures constitute the complete information to determine the dipole field with both its poles and the anisotropy, i.e. the corresponding velocity.

The temperature of the background radiation constitutes a real function consisting of a monopole and a dipole function on the celestial sphere, according to Eq. (7.45). For a given coordinate system (θ, φ), one pole of the dipole field is at point $P(\theta_p / \varphi_p)$, which is the apex of the motion, the second pole diametrically at $Q(\pi - \theta_p / \pi - \varphi_p)$. The dipole function must be cylindrically symmetric about the axis PQ, i.e. the numerical values of the function are constant along any circle in a normal plane to the axis. The same applies to the angle between the axis and the vectors from the center O of the sphere to a point R on such a circle, which can be expressed by the scalar product of the vector $\mathbf{p}$ pointing from point O to point P and the vector $\mathbf{r}$ pointing from point O to point R. Again, a Cartesian coordinate system is defined with the z-axis pointing towards the pole of the spherical coordinates, $\theta = 0$, and the x-axis towards $\varphi = 0$. In these coordinates, the scalar product of the vectors $\mathbf{p}$ and $\mathbf{r}$ is

$$\begin{aligned}
\mathbf{p}\cdot\mathbf{r} &= \begin{pmatrix} \sin\theta_p\,\cos\varphi_p \\ \sin\theta_p\,\sin\varphi_p \\ \cos\theta_p \end{pmatrix} \cdot \begin{pmatrix} \sin\theta\,\cos\varphi \\ \sin\theta\,\sin\varphi \\ \cos\theta \end{pmatrix} \\
&= \sin\theta_p\,\cos\varphi_p\,\sin\theta\,\cos\varphi + \sin\theta_p\,\sin\varphi_p\,\sin\theta\,\sin\varphi + \cos\theta_p\,\cos\theta \\
&= \cos\theta_p\,\cos\theta + \sin\theta_p\,\sin\theta\,\cos(\varphi - \varphi_p) = \cos\theta_r\,.
\end{aligned} \tag{7.50}$$

Since $\cos\theta_r$ is a constant on a circle in a plane normal to the axis OP, this function is a proportion of a dipole function, as required in Eq. (7.45),

$$\frac{1}{T'} = \frac{1}{T_{\text{av}}}\,\gamma\,(1 - \beta\,\cos\theta_r)\,. \tag{7.51}$$

Indeed, 4 observed temperatures at 4 points yield 4 equations,

$$\frac{1}{T_i'} = \frac{\gamma}{T_{\text{av}}}\,\{1 - \beta\,[\cos\theta_p\,\cos\theta_i + \sin\theta_p\,\sin\theta_i\,\cos(\varphi_i - \varphi_p)]\} \qquad \text{for } i = 1, 2, 3, 4 \tag{7.52}$$

for the 4 unknowns T_{av}, β, θ_p and φ_p. This is a system of 4 non-linear equations that must be solved by some adequate numerical methods.

Encounter

The event where the world lines of the two spacecrafts cross, is an encounter. This event realizes the scenario of the encounter of two local observers (Section 1.3.1). The local observers or astronauts are considered to move through space in spacecrafts, from which they can map the celestial sphere, that is, they can locate visible stars, and galaxies, by the direction in a spherical coordinate system. Let us suppose that spacecraft O_s meets a second spacecraft O_t at one given moment in time. Both astronauts map their celestial spheres at that moment and later exchange these maps by some means of radio transmission. In an idealized case, this encounter shall be considered a coincidence at one point in space and time, or event in spacetime.

If the direction of the relative motion between the two local observers at the moment of coincidence is known, the conformal transformation can easily be determined by the directions of one single star with Eq. (2.15). If the observers have no information on the direction of their relative motion at the moment of coincidence, the conformal mapping is not obvious. In general, a linear fractional transformation is defined by three points and their images. In the case of the aberration, we seek a map corresponding to Eq. (2.4) and the situation illustrated in Figure 2.3. Thus, the two spheres must be rotated with respect to each other in such a way that all great circles through the points and their corresponding images meet in two diametrical points on the sphere which are the apex and antapex of the motion of e.g. observer O_s.

The two observers coincide according to the scenario defined in Figure 2.3. They agree on 3 identifiable stars and map these stars on their individual spherical coordinate systems (Riemann spheres):

- Observer O_s maps the stars S_1, S_2 and S_3 given by the complex numbers $\mathfrak{s}_1$, $\mathfrak{s}_2$ and $\mathfrak{s}_3$ on his Riemann sphere, and
- Observer O_t maps the stars T_1, T_2 and T_3 given by the complex numbers $\mathfrak{t}_1$, $\mathfrak{t}_2$ and $\mathfrak{t}_3$ on his Riemann sphere.

The apex A is defined by the direction of the motion of observer O_s with respect to observer O_t and is given by the complex numbers $\mathfrak{a}_t$ and $\mathfrak{a}_s$ in the respective coordinates of O_t and O_s. The given problem is the determination of $\mathfrak{a}_s$ and $\mathfrak{a}_t$.

Here we only outline the solution procedure without going into the details of the mathematics and the solution methods for the resulting equations. The following geometrical operations can be performed:

- Rotation R_s: rotate the sphere of O_s such that apex A coincides with the origin $\mathfrak{z}$=0 of the Riemannian sphere, $\mathfrak{a}_s \to 0$, and $(\mathfrak{s}_1, \mathfrak{s}_2, \mathfrak{s}_3) \to (\bar{\mathfrak{s}}_1, \bar{\mathfrak{s}}_2, \bar{\mathfrak{s}}_3)$.
- Rotation R_t: rotate the sphere of O_t such that apex A' coincides with the origin $\mathfrak{z}$=0 of his Riemannian sphere, $\mathfrak{a}_t \to 0$, and $(\mathfrak{t}_1, \mathfrak{t}_2, \mathfrak{t}_3) \to (\bar{\mathfrak{t}}_1, \bar{\mathfrak{t}}_2, \bar{\mathfrak{t}}_3)$,
- Rotation-expansion D: find a rotation-expansion $D(\mathfrak{z}) = \mathfrak{d}\,\mathfrak{z}$ with $\mathfrak{d} \in \mathbb{C}$ such that $(\bar{\mathfrak{s}}_1, \bar{\mathfrak{s}}_2, \bar{\mathfrak{s}}_3) \to (\bar{\mathfrak{t}}_1, \bar{\mathfrak{t}}_2, \bar{\mathfrak{t}}_3)$

The two triples of coordinates of the stars uniquely define a linear fractional map such that

- C: $(\mathfrak{s}_1, \mathfrak{s}_2, \mathfrak{s}_3) \rightarrow (\mathfrak{t}_1, \mathfrak{t}_2, \mathfrak{t}_3)$

Map C can be composed of the following consecutive maps

$$R_t^{-1} \cdot D \cdot R_S = C\,. \tag{7.53}$$

Map C: except for some ill-conditioned configurations, any 3 points and their corresponding images define a conformal mapping of the celestial sphere, given by a linear fractional transformation of the Gaussian sphere. In principle, the three points and their maps could be substituted into Eq. (2.4) and solve the three non-linear equations by some numerical scheme. Mappings of three points on some carefully chosen special points may be determined by analytical methods, e.g. mapping one of the points to the point $\mathfrak{z}_1$=0, a second point to $\mathfrak{z}_2$=1, and the third point to infinity, $\mathfrak{z}_3=\infty$.

Rotations: the general form of a rotation of the Riemann sphere was proposed by Gauss in 1819 (Needham, 1997),

$$R(\mathfrak{z}) = \frac{\mathfrak{a}\mathfrak{z} + \mathfrak{b}}{-\bar{\mathfrak{b}}\mathfrak{z} + \bar{\mathfrak{a}}}\,, \tag{7.54}$$

where $\bar{\mathfrak{a}}$ and $\bar{\mathfrak{b}}$ are the conjugate complex of the coefficients $\mathfrak{a}$ and $\mathfrak{b}$, respectively. In the given case, the rotations are special since a point P given by the complex number $\mathfrak{p}$ is rotated into the origin $\mathfrak{z} = 0$ of the sphere, thus

$$R(\mathfrak{z}) = \frac{\mathfrak{z} - \mathfrak{p}}{\mathfrak{p}\mathfrak{z} + 1}\,, \tag{7.55}$$

or vice versa.

Rotation-expansion: the composition of the rotations and the aberration (rotation-expansion), Eq. (7.53), gives three equations for the three complex unknowns $\mathfrak{a}_\mathfrak{s}$, $\mathfrak{a}_\mathfrak{t}$, and $\mathfrak{q}$ (6 equations for the 3 real and 3 imaginary parts of the unknowns),

$$\begin{pmatrix} 1 & -\mathfrak{a}_\mathfrak{s} \\ \bar{\mathfrak{a}}_\mathfrak{s} & 1 \end{pmatrix} \cdot \begin{pmatrix} 1 & 0 \\ 0 & \mathfrak{q}^{-1} \end{pmatrix} \cdot \begin{pmatrix} 1 & -\mathfrak{a}_\mathfrak{t} \\ -\bar{\mathfrak{a}}_\mathfrak{t} & 1 \end{pmatrix}^{-1} = \begin{pmatrix} 1 & \mathfrak{r}_\mathfrak{b} \\ \mathfrak{r}_\mathfrak{c} & \mathfrak{r}_\mathfrak{d} \end{pmatrix}. \tag{7.56}$$

Note that we have normalized the transformation matrix such that the coefficient in the upper left corner is one. The solution gives the apex with respect to the second spacecraft only. If this spacecraft is very slow the solution gives an approximate information of the apex with respect to the cosmic background frame.

7.3.2 Navigation

Navigation is the process of monitoring and controlling the movement of a space craft or vehicle from one place to another. If a map is available, in our case a 4-dimensional map of the stars or galaxies, and if we can identify our position and directions on the map, then we can navigate from one point in space to another. Notably navigation also asks for manoeuvrability, which is limited at hyperrelativistic speeds of the spacecraft with respect to the cosmic background.

Observing the sky

In the previous section, the orientation with respect to frame of the cosmic background radiation or with respect to another spacecraft is described. Yet, there are limits to such orientation procedures that are closely related to the relativistic effects of aberration and Doppler shift at hyperrelativistic velocities (Lagoute and Davoust, 1995).

With increasing velocity, the celestial sky is concentrated more and more into a small circle around the apex of the motion. On one side, this makes it easier to know the direction of the flight since e.g. the cosmic background radiation becomes brighter, shifts into the electromagnetic spectrum of higher frequencies and is easier to detect and localize. On the other hand, all of the visible universe is concentrated into the small circle around the apex, and the relative resolution decreases. Finally, we know very accurately the direction of motion but loose information on the direction relative to the galaxies.

The angular resolution of an optical system such as a telescope with lenses or mirrors is proportional to the wavelength λ of the observed signals and inversely proportional to the aperture $2R$ of the system (diameter of the lens). The Raleigh criterion states

$$\Delta\theta \approx 1.22\,\frac{\lambda}{2R}\,, \tag{7.57}$$

where $\Delta\theta$ is the smallest angle that can be resolved and 1.22 is a proportionality constant from diffraction theory. Let us make the (unrealistic) assumption that a civilization capable of constructing a Bussard ramjet as discussed above, is also able to construct a telescope with a given aperture for electromagnetic waves of a very large range of wavelengths.

Consider the following scenario for orientation in space: due to aberration, Eq. (2.15), the sky in the half-sphere with an opening angle $2\theta = \pi$ around the apex as seen by an observer at rest to the cosmic background radiation is mapped into a small cone around the apex of relative motion with an opening angle $2\theta'$, where

$$\tan\frac{\theta'}{2} \approx \frac{\theta'}{2} \tag{7.58}$$

applies we find

$$2\theta' \approx 2\sqrt{\frac{1-\beta}{1+\beta}} \approx 4\gamma, \tag{7.59}$$

If we want to resolve details in the small cone we divide the diameter of the mapped circle by $2n$ pixels and get with the angular resolution

$$\Delta\theta' = \frac{\theta'}{n}, \tag{7.60}$$

the required aperture of a telescope for a given wavelength for a moving observer

$$2R \approx 1.22\frac{n\,\lambda'}{\theta'}. \tag{7.61}$$

With the appropriate aperture and wavelength, the sky around the apex could be observed with any desired resolution at any velocity. As the cone $2\theta'$ shrinks with increasing relativistic factor γ, the wavelength λ' of the star-light decreases with γ. According to Eq. (2.24) the Doppler shift is

$$\lambda' = \lambda\sqrt{\frac{1-\beta}{1+\beta}} \approx \lambda\,2\gamma\,. \tag{7.62}$$

As the cone of observation shrinks with the Doppler factor, Eq. (7.59), the resolution with a given aperture increases with the same factor due to the Doppler shifted wavelength. Thus, for a given aperture, the resolution remains independent of the velocity. This encouraging argument comes with a caveat. During the uniformly accelerated flight, the required wavelength eventually drops below 10^{-18} m that translates to a photon energy in the order of TeV. As such photons have a very low interaction cross section with baryonic matter and a detector would have to be very massive, at least. Thus, at extreme hyperrelativistic velocities, we would have to fly blind i.e. without distinction of objects across the universe.

Changing direction

Navigation may require a change in direction of motion in order to avoid unwanted collisions. To achieve this, at least part of the thrust of the spacecraft must be diverted in the transverse direction with respect to the momentary flight direction.

The smallest achievable radius of the curved trajectory is limited by the given velocity and the achievable thrust. It is not clear how transverse thrust could be performed technically with a Bussard ramjet. Though for some estimation of possible curved trajectories, we assume that the full thrust can be directed transversely without loss in forward velocity. This assumption gives the limit circle radius that can be flown at a given velocity.

Let a be the possible proper acceleration of the spacecraft and ω the momentary proper velocity at the time, when the thrust is diverted transversely. According to Eq. (3.30) and Eq. (3.16),

$$r = c\,\frac{\omega^2}{a} = c\,\frac{\sinh^2(a\tau)}{a}, \tag{7.63}$$

where r is the radius of the curve and τ the proper time elapsed since the start of the ramjet. Figure 7.5 shows the radius of the curve flown as a function of the distance covered since the start of the spacecraft Eq. (7.3) accelerating with $c\,\alpha = 1$ ly/a^2. The turning radius grows with the square of the covered distance and reaches the size of galaxies after a flight of less than one year.

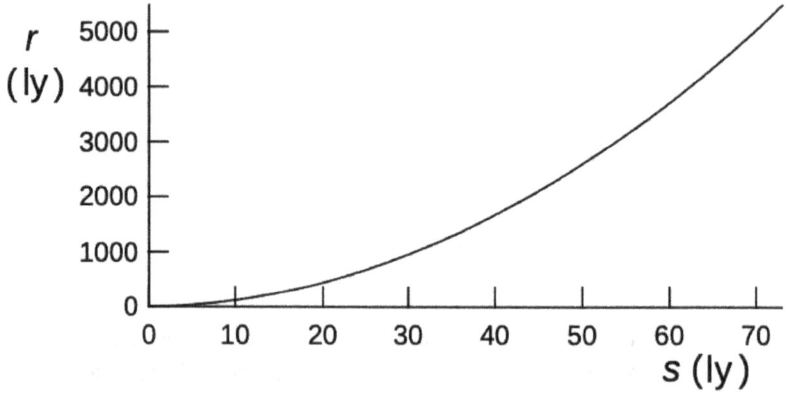

Fig. 7.5: Radius r of the curve flown as a function of the distance s covered since the start of a spacecraft accelerating with $c\,\alpha = 1$ ly/a^2.

In the coordinate system of the spacecraft the proper angular velocity (ds/dτ)/r of the rotation of the flight direction is,

$$\Omega_\mathrm{p} = \sqrt{\frac{c\,\alpha}{r}} = \frac{\alpha}{\omega}\,. \tag{7.64}$$

In the coordinate system of the Earth the angular velocity (ds/dt)/r of the spacecraft is with the time dilation dt=$\gamma\,$dτ, Eq. (4.46),

$$\Omega_\mathrm{E} = \frac{1}{\gamma}\sqrt{\frac{c\,\alpha}{r}} = \frac{\alpha}{\gamma\,\omega}\,. \tag{7.65}$$

The factor between the angular velocity as registered in the spacecraft Ω_p and that on Earth Ω_E is a direct result of relativistic kinetics and dynamics in circular motion as discussed in Section 3.2. The orbit times T_p and T_E are determined by the corresponding integrals

$$T_{\mathrm{p,E}} = \int_0^{2\pi} \frac{1}{\Omega}\,\mathrm{d}\varphi\,, \tag{7.66}$$

which yields $T_\mathrm{E} = \gamma\,T_\mathrm{p}$. This is another possibility for explaining the so called twin paradox (Section 4.6), where the accelerated twin in the spacecraft ages slower than the twin at rest. Importantly, the ability to fly a curve decreases with increasing velocity. To fly a curve with relativistic spacecraft encounters another serious limitation. If there is friction (Section 7.4) due to interstellar media the diversion of thrust takes forward linear momentum from the spacecraft. Therefore, to fly a curve without loss of forward velocity is only possible if only part of the thrust is diverted. This fraction decreases with increasing velocity as the friction of the interstellar media increases with increasing velocity.

7.3.3 Communication

Communication in spaceflight is exclusively performed by electromagnetic waves. With increasing distance, the time lag between the emission and reception of signals increases, and with increasing relative velocity, the frequencies shift. At hyperrelativistic velocities, both effects may rapidly become an obstacle that makes communication extremely difficult if not impossible. In addition, some regions of spacetime behind the spacecraft drop behind a horizon, from which no signal can reach the spacecraft anymore (Section 7.1.2).

Return trip to α Centauri

In Section 4.6, the treatment of the space-traveling twins is restricted to flights at uniform velocities. For this scenario, space traveling triples are needed because the abrupt change between the reference frames is not possible. With accelerated flights, one twin can in principle travel forth to a destination in space and return to Earth again. This twin is always accelerating while the other remains in the same inertial frame for the entire time span. When they meet again on Earth, they can compare their proper times.

Tab. 7.1: The trip to α Centauri (4.3 light-years distance) and back with proper accelerations of $c\,\alpha$=g as outlined in Figure 7.6: the first and second row give the ship proper times and the coordinate times when the ship is at the corresponding points of the journey. The third row gives the arrival coordinate time when a signal sent from the ship at the corresponding points arrives at Earth. All times given in years.

	M: midway to α Centauri	A: arrival at α Centauri	N: midway from α Centauri	R: arrival at Earth
ship proper time	1.82	3.63	5.45	7.26
coordinate time	2.99	5.98	8.96	11.94
signal at Earth	5.14	10.28	11.11	11.94

In Figure 7.6 the flight proposed in Section 7.1.2 to α Centauri and back is outlined. The astronaut-twin accelerates or decelerates with $c\,\alpha = 1\,\text{ly/a}^2$ and thus reaches a peak velocity of $\beta_p = 0.88$ at the midpoint between Earth and α Centauri. Let the twins transmit

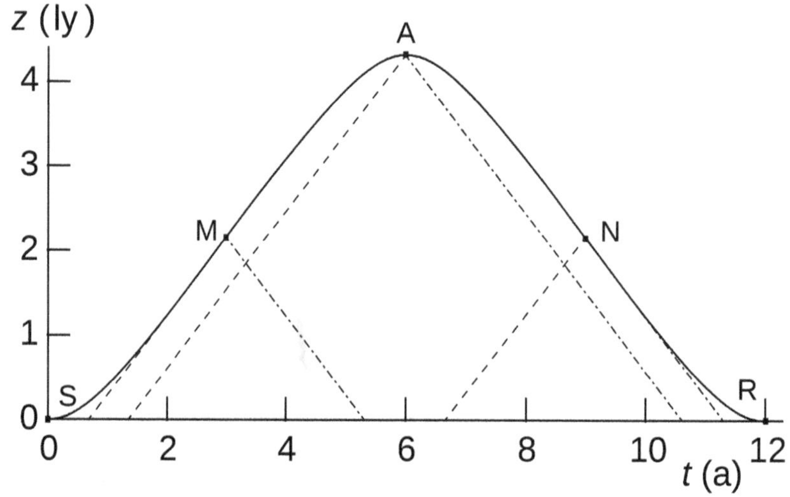

Fig. 7.6: The flight to *α* Centauri. The four legs (SM, MA, AN and NR) of the journey correspond to congruent hyperbolic space-time diagrams. The dashed lines denote the space-time lines of the signals sent from Earth to the ship and the dash-dotted lines from the ship to Earth. Explanations and numerical values are given in the text and in Table 7.1

their own proper-time signals with the speed of light, and the corresponding other twin records the arrival time of the signals in his own proper time scale. Müller et al. (2008) discuss in detail the continuous observation of such time signals. To simplify the mathematics, we restrict the computation of the proper times of the reception of the signals to the special points midway through the trip and to the arrivals at *α* Centauri and Earth. For simplicity, we call the starting point at Earth S, the arrival point at *α* Centauri A, the midway points M and N for the trip to *α* Centauri and back, respectively, and the final return to Earth point R (Table 7.1).

Tau Zero

The novel *Tau Zero* (Anderson, 1970) tells the story of the crew in a Bussard ramjet. They planned to go to the third planet of Beta Virginis about 35.6 light-years from Earth. Shortly after starting to operate the ramjet or "starship Leonora Christine", the encounter with a dust cloud damaged the break of the engine. To make repairs, the ramjet engine had to be stopped. The rain of energetic particles and the strong radiation (see Section 7.4) at the already relativistic velocity would have damaged the spacecraft and killed the crew and therefore the group of 100 people in the ship was forced to continue the accelerated journey, seemingly forever.

The dynamics of hyperrelativistic propulsion of of a spacecraft has been discussed in Section 7.2.4 and by Blatter and Greber (2017) before. For the communication in the "Tau Zero scenario" consider a spacecraft that starts from Earth at time $t = 0$ and accelerates with constant proper acceleration $c\,\alpha = 1\,\text{ly/a}^2$ (corresponding to about 1 g) along the z direction. Signals sent from Earth at time t_1 after the start of the spacecraft reach a distance

$$z_1 = c\,(t - t_1) \tag{7.67}$$

from Earth at coordinate time t. The spacecraft reaches a distance z_s at coordinate time t_s according to Eq. (7.6). Setting $z_1 = z_s$ yields the time when the signal reaches the ship,

$$t = t_1\,\frac{2 - \alpha\,t_1}{1 - \alpha\,t_1}\,. \tag{7.68}$$

According to the communication horizon behind the ship (Figure 7.1), this function has a singularity at $z_1 = \alpha\,t_1$, thus the signals arrive at a later and later time the closer the signal was sent before t_1. The spacecraft outruns all signals sent at or later than t_1. Figure 7.7 shows the distance of the ship z_s when it receives a signal as a function of the time t_1, when the signal was sent from Earth.

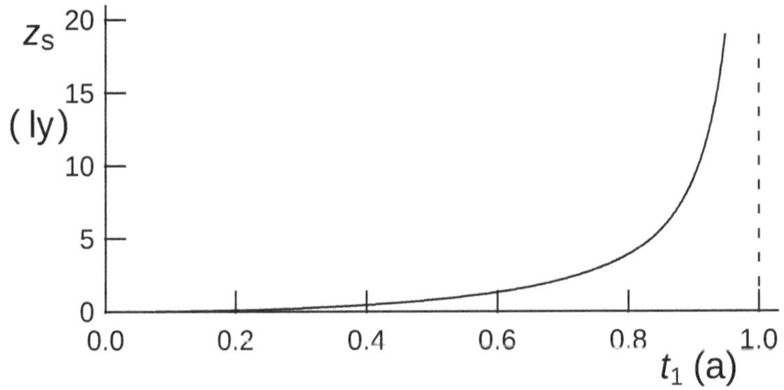

Fig. 7.7: Distance z_s (in light-years) at which a spacecraft accelerating with $c\,\alpha = 1$ (in light-years per year^2) receives a signal from Earth at coordinate time t_1.

Reversely, signals sent from the ship can always reach Earth in a finite time,

$$t_r = t_s + \frac{z_s}{c}\,, \tag{7.69}$$

where t_s and t_r are the coordinate time of the transmission from the ship and reception on Earth of the signal and z_s is the distance of the ship at t_s from Earth. Signals sent out after a few years of proper ship time arrive many thousand years later in Earth coordinate time.

Closely related to the increasing delay of the signals sent from the ship to Earth is the Doppler redshift of the signal to lower frequencies. At both ends of the electromagnetic frequency spectrum, there are some limiting factors for the transmission and reception of signals. The one-way communication after one year in the "Tau Zero scenario" reduces the problem to large Earth based antennas for the reception of signals from the the starship Leonora Christine, and to collimated high energy photon senders on the starship. On the low-frequency side, the size of antennas needs to be very large to receive signals efficiently. On the other hand, high frequencies with energetic photons cause high power budgets on the starship.

Eventually, the spacecraft will not be able to communicate with the rest of the universe in which stars and galaxies generally move with moderately relativistic velocities with respect to the rest frame of the cosmic background radiation. Due to aberration, most of the universe is concentrated in a small blue-shifted area near the apex of the motion and almost no targets are left in the domain of directions of redshift. The fate of the spacecraft that is constantly accelerating in one direction is not very attractive: eventually the crew will not be able to communicate with the rest of the universe nor will it be able to observe anything in the rest of the universe. They will be completely isolated and alone, even in an universe densely populated with civilizations ready to communicate (Desloge and Philpott, 1987; Lagoute and Davoust, 1995; Müller et al., 2008).

The constantly accelerated spacecraft in the novel Tau Zero can in principle fly to cosmic distances within the lifetime of humans. The proper time on the starship becomes so much slower than the coordinate time such that a significant part of the evolution of the universe may pass. In the story, the universe finally begins to shrink and collapse again, a process often referred to as *big crunch*, opposite to the *big bang* at the beginning of the expansion. The crew of the spacecraft noticed the change from redshift of an expanding universe to a blue shift of a contracting universe. In view of their extreme hyperrelativistic velocity and resulting aberration and Doppler shift, it is questionable if they would have been able to observe and realize this change at all.

7.4 Horizons and obstacles

Interstellar and intergalactic space is a vacuum far emptier than the best vacuum achievable in laboratory conditions. Nevertheless, a relativistic spacecraft is exposed to a substantial drag due to particles and radiation that it encounters in flight. This will increase friction, and make the universe more and more opaque for any accelerating spacecraft. Eventually, the cosmic microwave background will be Doppler shifted to energies that enable photon-assisted proton decay (Greisen, 1966).

7.4.1 Cosmology and limits of special relativity

Astronomical (cosmological) observations, such as the redshift of the light from distant galaxies indicates an expansion of the universe. Although a mathematical description of the theory of gravitation (general relativity) and cosmology lies beyond the scope of this book, we want to outline qualitatively some results concerning cosmic expansion, for a comprehensive description, see Davis and Lineweaver (2001), Davis (2003) and Davis and Lineweaver (2004).

Edwin Hubble (1936) observed the red shifted light of galaxies and interpreted this fact as a result of the motion of the *nebula* (galaxies). The redshift can not be understood with the Doppler shift of receding sources in Eq. (2.19), but also involves general relativity. Despite the notoriously difficult task to estimate distances of galaxies he suggested a proportionality between the expansion rate $\dot{R}$ and the distance R of galaxies,

$$\dot{R} = H\,R\,, \tag{7.70}$$

where H is the *Hubble constant*. This relation holds at least in a statistical sense. The Andromeda galaxy for instance moves towards our Milky Way galaxy due to a gravitational bond and will collide with our Milky Way galaxy in about $5 \cdot 10^9$ years.

A Hubble type universe has some peculiar qualities. First, such an expansion does not qualify any point as the center of the expansion. This can be visualized by a rubber band that is stretched with a constant velocity at both ends. An observer in any point on the rubber band would see other points of the band to move away from him with velocities proportional to the distance. This equally applies to all points on the band. The same holds for the surface of an inflating balloon as for three-dimensional space.

A second feature is the fact that the expansion rate $\dot{R}$ reaches the speed of light at a distance

$$R_{\mathrm{H}} = \frac{c}{H} \tag{7.71}$$

from an observer, which is referred to as the *Hubble radius*, defining the so called *Hubble sphere* around the observer. This is not in conflict with special relativity. As stated above, any observer in a Hubble type universe observes the same situation. The observed motion of an object with velocity v_{r} can be split into two parts (Davis, 2003),

$$v_{\mathrm{r}} = v_{\mathrm{rec}} + v_{\mathrm{pec}}\,, \tag{7.72}$$

where v_{rec} is the Hubble recession velocity and v_{pec} is the peculiar velocity of the object relative to its local Hubble frame of reference. Like on Earth, the speed of light in any such frame is a peculiar velocity, $v_{\mathrm{pec}} = c$. Notably, the limitation of velocities in special relativity only applies to v_{pec}, but not to the Hubble recession velocity v_{rec}.

The question arises if the Hubble sphere constitutes the limits of the visible universe, or under what circumstances can light be observed even if it is emitted towards us from galaxies outside the Hubble sphere. This light first recedes from us although it was sent out towards us. Depending on the temporal evolution of Hubble's constant, the Hubble sphere may also expand, and eventually, this light enters the sphere and starts to propagate toward us and make the galaxy visible. Thus the Hubble sphere is not a *horizon*. For more details see Davis (2003).

These results stem from a cosmology based on general relativity for an isotropic, and homogeneous universe sometimes referred to as Friedmann-Robertson-Walker cosmology. Another consequence is the fact that the observed redshift of the light from distant

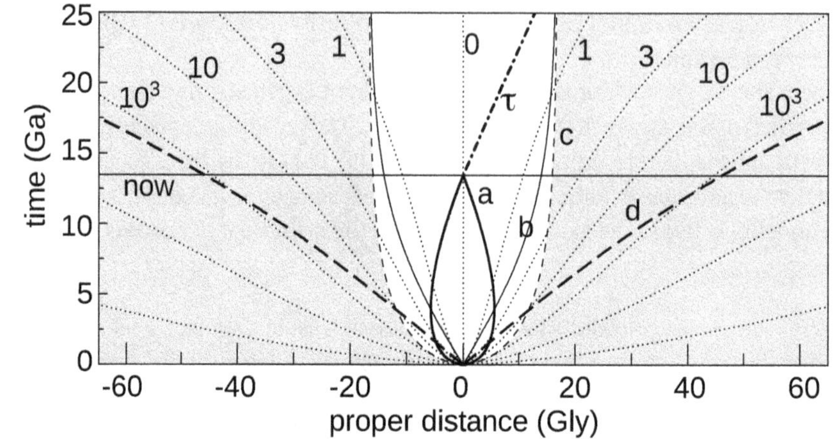

Fig. 7.8: Flight of a constantly accelerating spacecraft in an expanding universe with a Hubble's constant of 70 km s^{-1} Mpc^{-1}: spacetime diagram adapted from Davis and Lineweaver (2004). The dotted lines show world lines of co-moving objects, i.e. galaxies moving with the general expansion with the numbers indicating the current redshift. The thick solid line labeled a is the past light cone, the thin solid line labeled b the Hubble sphere, the thin dashed line labeled c is the event horizon (see text) and the thick dashed line labeled d is the particle horizon (see text). The world line of the spacecraft (thick dash-dotted line labeled τ closely follows the future light cone.

galaxies does not correspond to the special relativistic Doppler shift due to the velocity v_{pec}. As Eq. (2.19) shows, the Doppler shift in special relativity goes to infinity if the velocity approaches the speed of light. The observed cosmic redshift is a cosmological redshift due to the expansion of space and perhaps in some cases a combination of cosmic redshift due to v_{rec} and Doppler shift due to some v_{pec}.

A constantly accelerating Bussard ramjet will reach places with higher and higher recession velocities with respect to Earth, but its peculiar velocity will be hyperrelativistic and still increase closer and closer to the speed of light. Eventually, the spacecraft will reach Earth's Hubble sphere. This limits the validity of Eq. (7.69), which describes the arrival time of signals from the ship on Earth in Minkowski space, which is only valid as long as the distance to Earth is substantially smaller than the Hubble radius. This does not necessarily mean that signals sent from the ship beyond the Hubble sphere cannot eventually be received on Earth, although Earth likely will not exist anymore at the expected time of message arrival. Although the spacecraft travels at highly relativistic velocity with respect to its local Hubble frame of reference, the signals it sends back still travel with the speed of light with respect to this frame. Thus the observability of the spacecraft is in principle the same as the observability of galaxies with much smaller non-relativistic peculiar velocities. The difference to signals (light) from galaxies is the additional large relativistic Doppler shift due to the large peculiar velocity of the ship.

Figure 7.8 depicts a spacetime diagram of an expanding universe together with the world line of an accelerating spacecraft at high relativistic velocity. Except for the very beginning of the flight, the spacecraft follows closely the future light cone. It will reach the Hubble sphere in about 14 Gigayears coordinate time, almost independent of the acceleration. The proper time needed, on the other hand, depends on the acceleration. At an acceleration of 1 ly/a^2 the Earth-centered Hubble sphere will be reached in a few decades proper time. The event horizon, i.e. our past light cone at the end of time, $t = \infty$, constitutes a true horizon, no light can travel further than to the event horizon. The world line of the spacecraft crosses the Hubble sphere at a time when the event horizon is not much further away, thus shortly after it also crosses the event horizon (Figure 7.8). At the latest after this moment, the spacecraft is totally detached from Earth or at least the position where Earth once was. Earth already fell behind the kinematic horizon of the constantly accelerated spacecraft such that no signal from Earth can be received anymore in the spacecraft. Now, the spacecraft drops behind the event horizon such that also no signal from the spacecraft can reach Earth anymore.

7.4.2 Interstellar matter

Within galaxies the density of interstellar matter is estimated to be about one hydrogen atom per cubic centimeter. Between the galaxies, it is about 6 orders of magnitude smaller with about 1 hydrogen atom per cubic meter. Within dense nebulae, particles made of ice and carbon of 10^{-18} kg may occur (Spitzer, Jr., 1978), but they contribute only to about half a percent of the total mass.

As long as the Bussard ramjet is operational, interstellar matter is collected and funneled through the ship's reactor. Fusion then accelerates the hydrogen and ejects it through the rear of the spaceship. In this case, hydrogen acts as propellant and does not pose a danger to the ship. On the other hand, if the collection of the hydrogen stops due to some malfunction of the engines, the particles and atoms are stopped and collected by the ship. The momentum and energy transfer from the particles to the ship then decelerates and heats the ship.

To calculate the particle flux $\dot{N}_r$, the proper velocity ω is the adequate quantity. The number of collected particles depends on the volume $\mathrm{d}V$ of space covered by the spaceship as measured in the frame of reference of Earth, seen from the spacecraft, the flux must be measured in units of proper time $\mathrm{d}\tau$. The volume $\mathrm{d}V$ covered in a proper time interval $\mathrm{d}\tau$ by a spaceship with a cross sectional area A at a proper velocity ω is

$$\mathrm{d}V = A\omega c\,\mathrm{d}\tau\,, \tag{7.73}$$

and the corresponding particle flux

$$\dot{N}_r = A\omega c n\,, \tag{7.74}$$

where n is the density of particles. With the assumption that the majority of particles are hydrogen atoms with a mass of $\mu_r = 1.7 \cdot 10^{-27}$ kg, and with a particle density in intergalactic space of $n_{\text{intergalactic}} = 1\ \text{m}^{-3}$, the particle flux on the spaceship at a velocity of $\omega = 10^6$ becomes $\dot{N} = 3 \cdot 10^{14}\ \text{s}^{-1}\text{m}^{-2}$, and the corresponding mass flux is $5 \cdot 10^{-13}\ \text{kg}\,\text{s}^{-1}\text{m}^{-2}$.

The momentum p_r of a particle with mass μ_r is

$$p_r = \mu_r \omega c\,, \tag{7.75}$$

and thus, the pressure P on the front of the ship exerted by this particle flux increases with the square of the proper velocity,

$$P = \frac{\dot{N}_r p_r}{A} = \mu_r n \omega^2 c^2\,, \tag{7.76}$$

corresponding to 150 Pa for the above situation. The kinetic energy that each particle deposits in the ship is

$$E_{\text{kin}} = \mu_r c^2(\gamma - 1)\,. \tag{7.77}$$

For a proton this corresponds to about 1 TeV ($1.6 \cdot 10^{-7}$ J) at $\omega = 10^6$. This energy can be reached in today's most powerful accelerators, such as the Tevatron at Fermilab. The corresponding energy flux at the front of the ship is then $4.5 \cdot 10^{10}\ \text{Wm}^{-2}$, corresponding to a black body radiation with a temperature of nearly 30'000 K.

What happens if the ship hits a particle of ice or carbon with a mass of 10^{-18} kg or larger is not so clear. If the particle hits the shield, it may just go through the shield and possibly leave a hole in it. If the particle hits the spaceship, it may be stopped and deposes its entire momentum and energy in the ship. One particle with a mass of 10^{-18} kg and a velocity $\omega = 10^6$ deposits an energy of 10^{21} eV or about 100 J, which corresponds to the kinetic energy of a tennis ball with a speed of 215 km/h.

7.4.3 Electromagnetic radiation

The space ship flies through a field of electromagnetic radiation. The sources of this radiation are the stars and the cosmic microwave background, and its spectrum ranges from radio waves to gamma rays and beyond. The energy of the cosmic microwave background dominates the background radiation in deep space far away from stars (Sandage et al., 1993). At relativistic velocities the radiation field is strongly changed due to Doppler shift and aberration (Komar, 1965; Blatter and Greber, 1988). The astronaut registers a radiation field of high intensity and shifted to high frequencies around the apex, i.e. in the direction of the journey, and very little radiation from the rest of the celestial sphere.

The universe is mostly dark, except for the comic background radiation, which glows in the microwave spectrum. It is an almost perfect black body radiation with a temperature of 2.725 K, i.e. with a peak frequency of $\nu_0 = 2.82 \cdot 10^{11}\ \mathrm{s}^{-1}$. The density of photons in the universe is $4.11 \cdot 10^8\ \mathrm{m}^{-3}$. At high velocities (close to the speed of light), a spacecraft starts to feel the radiation pressure of the background radiation.

The number of photons that the ramjet collects in proper time $\Delta\tau$ is

$$N = A\, \varrho_i\, c\omega\, \Delta\tau\,, \tag{7.78}$$

where A is the area which the ramjet exposes to the radiation, ϱ_i the density of photons, c the speed of light, ω the proper velocity. The rate of collection of photons on $1\ \mathrm{m}^2$ is

$$\dot{N} = \varrho_i\, c\omega = \varrho_i\, c\gamma\,, \tag{7.79}$$

since at very large proper velocity, $\omega \approx \gamma$.

The momentum of one single photon with frequency ν is

$$p = \frac{h\,\nu}{c}\,, \tag{7.80}$$

where $h = 6.62607015 \cdot 10^{-34}\ \mathrm{m}^2\ \mathrm{kg\ s}^{-1}$ is the Planck constant. At very large velocities, the ramjet collects photons from a small area around the apex of the flight. These photons are Doppler shifted,

$$\nu = \nu_0\, 2\,\gamma\,, \tag{7.81}$$

thus, the total pressure of the photons is

$$P_{\mathrm{tot}} = 2\,h\,\nu_0\,\gamma^2\,\varrho_i\,. \tag{7.82}$$

To accelerate a ramjet with a mass $m = 10^8$ kg at $a = 10\ \mathrm{m\,s}^{-1}$ requires a force $F = m\,a = 10^9$ N. Thus, the maximum pressure the ramjet can take is $P = F/A$, where A is the collecting area of the ramjet. If $A = 10^{13}\ \mathrm{m}^2$ (Earth size), then $P_{\mathrm{max}} = 10^{-4}$ Pa (10^{-9} bar), and at a $\gamma \approx 10^4$ the entire thrust is used to just maintain this velocity. If the proton collecting funnel consists of graphene, which is almost perfectly transparent at this high frequency of photons, then the resulting pressure is much smaller, and the ramjet can accelerate to much higher velocities.

There are other processes that limit the achievable velocity of the ramjet, not because of its performance but because of its material, which starts to disintegrate due to UV and electromagnetic radiation with higher frequencies. Photons with an energy of about 5 eV (electron volt, $1\ \mathrm{eV} = 1.6 \cdot 10^{-19}$ J), i.e. with a wavelength of $\lambda = 2 \cdot 10^{-11}$ m or a frequency of about $\nu = 1.5 \cdot 10^{15}\ \mathrm{s}^{-1}$, which is in the ultraviolet range, start to damage the electronic bonds between atoms, and thus disintegrate the material. According to Eq. (7.81), the speed limit due to the cosmic microwave background is at $\gamma \approx 2.5 \cdot 10^3$.

Finally, there is a process that poses an absolute upper limit to the speed of a Bussard ramjet, unless the highly Doppler-shifted background radiation could be shielded.

This process also explains the high energy GZK cutoff of the protons in the cosmic radiation (Greisen, 1966; Zatsepin and Kuzmin, 1966; J. Abraham et al., 2008). Protons with a kinetic energy larger than $7 \cdot 10^{19}$ eV may react with the Doppler shifted photons of the cosmic background radiation into pions (Gaisser et al., 2016). With the proton mass $m_p = 1.67 \cdot 10^{-27}$ kg, and Eq. (3.39) for the kinetic energy, we obtain a relativistic factor $\gamma = 7.5 \cdot 10^{10}$, which sets a limit for the Bussard ramjet because at higher velocities, the cosmic background radiation starts to erode the spacecraft via pion production from protons and photons.

7.5 Conclusions

The last Chapter is kind of sobering and concerns the possibilities for space travel in spacecrafts. Rockets that exhaust material carried by the craft, such as chemical rockets or ion boosters, are limited in the achievable speed to non-relativistic speeds. The exhaust velocity may be increased but the required energy increases with the square of the exhaust velocity. Therefore, rockets can only be used in the vicinity of stars, but not really for interstellar flight. A Bussard ramjet, which collects interstellar matter for fuel, is limited in its achievable speed by the loss of momentum and energy during the collection and use of the collected matter. It is limited to perhaps 5-10% of the speed of light. Such velocities make a trip to α Centauri, the closest star, a venture that is longer than a human's life span. Even an ideal ramjet, is limited in the final velocity, mostly due to the interaction of the material of the ramjet with the cosmic background radiation.

Science fiction proposes several ways around the limitations of energy and momentum conservation, and the time problem. In the book *Islands of Space*, John W. Campbell (1930) first introduced warped space to circumvent special relativity, where the *Star Trek* movies (Roddenberry, 1979) use the name "wrap drive" for the same idea. The *Stargate* movie (Emmerich, 1994) assumes a network of gates, something like wormholes through space, to travel to stars in no time. Interstellar travel through "hyperspace", perhaps higher dimensions or multiverses, is proposed by Isaac Asimov in the Foundation series (Asimov, 1951) and by John Scalzi in the Old Man's War (Scalzi, 2005). So-called gravitic spacecrafts were suggested by Asimov, which take the energy from the vacuum to fly without inertial effects through space (Asimov, 1981). Although one may predict a very high vacuum energy, i.e. a vacuum catastrophe (Adler et al., 1995), the vacuum energy appears to be around zero and is thus not truly useful as an energy source for propulsion. Each of these suggestions requires new physics and technology, which is not visible even in rough hints. This means, likely we are restricted to our planetary system for the foreseeable future, and perhaps for a much longer time, if not forever.

Epilogue

What, if our world, like the world of Mister Tompkins, would be relativistic even in our daily life? What would that mean for the history of physics? Would this have delayed the development of physics because of the larger complexity of the phenomena, to begin with, or would this have accelerated the progress of physics because of the greater need to understand the complex world? The interdependence of space and time and motion would possibly be more intuitive. On the other hand, keeping track of the world lines of relatives and friends would become very difficult and the age structure of a group of people, e.g. family or school classes, could change from meeting to meeting. This might make social life and social rules more complex and difficult than they already are in our *non-relativistic* world. Lucky us that the speed of light is so large.

The existence of electromagnetism, electricity, and its applications in daily life, makes our world a *relativistic* world, yet in a very abstract way in phenomena such as light, magnets, motors, and other electronic devices. In electrodynamics relativity is only visible through a theoretical construction, which for *ordinary* people (say non-physicists) is certainly less intuitive.

The hypotheses of Komar (1965) are the core of this book. The isomorphism between the algebraic group of the conformal maps of the surface of a sphere, and the group of the proper Lorentz transformations is, according to late Res Jost, former Professor for theoretical physics at ETH Zurich (Jost, 1976):

> die Grundlage der Diracschen Theorie des Elektrons und ihre Faszinationskraft hat sie bis heute nicht verloren. Sie bleibt eine Quelle von Sinn und Unsinn.
> *the basis of Dirac's theory of the electron and to present, it has not lost its fascination. It remains a source of sense and nonsense.*

The consequent distinction and application of the three types of speeds in teaching Special Relativity allows us to explain the relativistic kinematics and dynamics in a more intuitive way. The description of motion, acceleration and linear momentum can be recovered in the classical way by replacing the coordinate velocity with the proper velocity. This fact is not novel since in vector notation, the spatial parts of velocity and acceleration vectors correspond to proper velocity and proper acceleration. But this is often obscured by taking the relativistic factor γ out of the vector components, thus writing the vectors as the relativistic factor times a vector with the components of the coordinate velocity.

We used the facts that e.g. an astronaut in a spaceship could not perceive a uniform motion of his spaceship without looking out of a window, and he would perceive a constant proper acceleration by feeling his own constant weight independent, however, of his velocity relative to any external frame of reference. These situations can be mapped to everyday experiences in traveling by trains or airplanes, and thus, can be based on the classical concepts of motion and inertia. The application of the relativistic kinemat-

https://doi.org/10.1515/9783111503592-009

ics and dynamics to steadily accelerated spaceflight may make the topic more appealing to students than misleading paradoxa and difficult concepts such as length contraction and time dilation.

The restriction to the two-dimensional 1+1 spacetime, i.e. the temporal and one spatial coordinate limits the description to collinear motion, acceleration and forces. The results offer the basis for the vector notation in two dimensions, which then can be extended to the 1+3 dimensional case. This predicts what an astronaut observes if he looks out of the window of his spacecraft, not length contraction, but aberration and Doppler shift of electromagnetic waves, and it is possible to compare the different time intervals that passed in the spacecraft and on Earth when she returns to Earth.

Bibliography

J. Abraham et al. Observation of the Suppression of the Flux of Cosmic Rays above 4×10^{19} eV. *Phys. Rev. Lett.*, 101(17):061101-7, 2008.

W. S. Adams. The Relativity Displacement of the Spectral Lines in the Companion of Sirius. *Proc. National Academy Sci. USA*, 11(7):382–387, 1925.

R.J. Adler, B. Casey, and O.C. Jacob. Vacuum catastrophe - an elementary exposition of the cosmological constant problem. *Am. J. Phys.*, 63(7):620–626, 1995.

T. Ahrens. Aether concept versus special relativity. *Am. J. Phys.*, 30(1):34–36, 1962.

H. Amar. New Geometric Representation of the Lorentz Transformation. *Am. J. Phys.*, 23(8):487–489, 1955.

P. Anderson. *Tau Zero*. Orion Publishing Group, London, 1970.

R. S. J. Anderson and G. E. Stedman. Distance and the conventionality of simultaneity in special relativity. *Found. Phys. Lett.*, 5:199–200, 1992.

R. S. J. Anderson and G. E. Stedman. Spatial measurements in special relativity do not empirically determine simultaneity relations: a reply to Coleman and Korté. *Found. Phys. Lett.*, 7(3):273–283, 1994.

N. Ashby. Relativity in the Global Positioning System. *Physics Today*, 55(5):41–47, 2002.

N. Ashby. Relativity in the Global Positioning System. *Living Rev. Relativity*, 6:1–42, 2003.

I. Asimov. *Foundation*. Bantam, New York, 1951.

I. Asimov. *Foundation's Edge*. Bantam. New York, 1981.

R. A. Bachman. Relativistic acoustic Doppler effect. *Am. J. Phys.*, 50(9):816–818, 1982.

C. G. Barkla. Secondary radiation from gases subject to X-rays. *The London, Edinburgh, and Dublin Philosophical Magazine and Journal of Science*, 373(30):685–698, 1903. doi:10.1080/147864403094629.

R. Beig and J. M. Heinzle. Relativistic aberration for accelerating observers. *Am. J. Phys.*, 76(7):663–670, 2008.

E. A. Bering, F. R. Chang-Díaz, J. P. Squire, V. Jacobson, L. D. Cassady, and M. Brukardt. High Power Ion Cyclotron Heating In the VASIMR Engine. In *45th AIAA Aerospace Sciences Meeting and Exhibit AIAA-2007-586*. American Institute of Aeronautics and Astronautics, Reno, Nevada, 2007.

R. A. Beth. Direct detection of the angular momentum of light. *Phys. Rev.*, 48(5):471, 1935.

H. Blatter and T. Greber. Aberration and Doppler shift: An uncommon way to relativity. *Am. J. Phys.*, 56(7): 333–338, 1988.

H. Blatter and T. Greber. Tau Zero: In the cockpit of a Bussard ramjet. *Am. J. Phys.*, 85(12):915–920, 2017.

M. Boezio and 33 others. Measurement of the flux of atmospheric muons with the CAPRICE94 apparatus. *Phys. Rev. D*, 62:032007, 2000.

M. Born. *Die Relativitätstheorie Einsteins*. Springer, Berlin, 1920.

J. Bradley. A Letter to Dr. Edmond Halley Astronom. Reg. c. Giving an Account of a New Discovered Motion of the Fix'd Stars. *Phil. Trans.*, pages 637–661, 1727.

B. D. Bramson. A Derivation of the Lorentz Transformation Assuming It to be Once Differentiable. *Am. J. Phys.*, 36(12):1163–1165, 1968.

R. W. Brehme. A Geometric Representation of Galilean and Lorentz Transformations. *Am. J. Phys.*, 30:489–496, 1961.

R. W. Brehme. The advantage of teaching relativity with four-vectors. *Am. J. Phys.*, 36(10):896–901, 1968.

R. W. Brehme. On the physical reality of the isotropic speed of light. *Am. J. Phys.*, 56(9):811–813, 1988.

J. R. Brophy and 21 others. Ion Propulsion System (NSTAR) DS1 Technology Validation Report. Technical Report 00-01, 10/2000, Jet Propulsion Laboratories, California Institute of Technology, 2000.

E. Buckingham. The principle of similitude. *Nature*, 96:396–397, 1915.

R. W. Bussard. Galactic matter and interstellar flight. *Astron. Acta*, 6:179–194, 1960.

J. W. Campbell. *Islands of Space*. AceBooks, INC., New York, 1930.

H. Cavendish. *The Scientific Papers of the Honourable Henry Cavendish, F.R.S.*, volume 2. Edited by E. Thorpe Cambridge University Press, London, 1921, 1784.

https://doi.org/10.1515/9783111503592-010

I. B. Cohen. *Römer and the First Determination of the Velocity of Light*. The Burndy Library INC, New York, 1944.

R. Coleman and H. Korté. On Attempts to Rescue the Conventionality Thesis of Distant Simultaneity in STR. *Found. Phys. Lett.*, 5(6):535–571, 1992.

A. H. Compton. A quantum theory of the scattering of x-rays by light elements. *Phys. Rev.*, 21(5):483–502, 1923.

C. A. Coulomb. Premier mémoire sur l' électricité et le magnétisme. *Histoire de l' Académie Royale des Sciences*, pages 569–577, 1785a.

C. A. Coulomb. Second mémoire sur lélectricité et le magnétisme. *Histoire de l'Académie Royale des Sciences*, pages 578–611, 1785b.

D. P. Cox and R. J. Reynolds. The Local Interstellar Medium. *Ann. Rev. Astron. Astrophys.*, 25:303–344, 1987.

D. Davis, W. F. Channing, and J. Jr. Bacon. *Manual of Magnetism*. D. Davis, Magnetical Instrument Maker, Boston, 1842.

T. M. Davis. *Fundamental Aspects of the Expansion of the Universe and Cosmic Horizons*. PhD thesis, The University of New South Wales, Sydney, Australia, 2003. arXiv:astro-ph/0402278v1.

T. M. Davis and C. H. Lineweaver. Superluminal recession velocities. *AIP Conference Proceedings*, 555:348–351, 2001.

T. M. Davis and C. H. Lineweaver. Expanding confusion: Common misconceptions of cosmological horizons and the superluminal expansion of the universe. *Publ. Astron. Soc. Australia*, 21(1):97–109, 2004.

E. A. Desloge and R. J. Philpott. Uniformly accelerated reference frames in special relativity. *Am. J. Phys.*, 55 (3):252–261, 1987.

P. A. M. Dirac. The quantum theory of the electron. *Proc. R. Soc. Lond.*, 117:610–624, 1928.

C. Doppler. Über das farbige Licht der Doppelsterne und einiger anderer Gestirne des Himmels. *Abhandlungen der Böhmischen Gesellschaft der Wissenschaften*, pages 456–482, 1842.

J. J. Dykla. Doppler redshift in oblique approach of source and observer. *Am. J. Phys.*, 47(4):381–382, 1979.

A. S. Eddington. On the Relation between the Masses and Luminosity of the Stars. *Monthly Notices Royal Astron. Soc.*, 34:308–359, 1924.

H. Ehrlichson. Comment on "A Unidirectional Test of Special Relativity". *Am. J. Phys.*, 41(11):1298–1299, 1973.

A. Einstein. Zur Elektrodynamik bewegter Körper. *Annalen Phys.*, 17(10):891–921, 1905a.

A. Einstein. Über einen die Erzeugung und Verwandlung des Lichtes betreffenden heuristischen Gesichtspunkt. *Annalen Phys.*, 17(8):132–148, 1905b.

A. Einstein. Relativitätsprinzip und die aus demselben gezogenen Folgerungen. *Jahrbuch der Radioaktivität*, 4:411–462, 1907.

A. Einstein. Über den Einfluss der Schwerkraft auf die Ausbreitung des Lichtes. *Ann. Phys.*, 35:898–908, 1911.

A. Einstein. Die Grundlage der allgemeinen Relativitätstheorie. *Ann. Phys.*, 49(4):770–822, 1916.

A. Einstein. How I created the theory of relativity: talk given in Kyoto on 14 December 1922. *Physics Today*, 8: 45–47, 1982.

R. Emmerich. Stargate (Film). Metro-Goldwyn-Mayer, 1994.

K. M. Evenson, J. S. Wells, F. R. Petersen, B. L. Danielson, G. W. Day, R. L. Barger, and J. L Hall. Speed of Light from Direct Frequency and Wavelength Measurements of the Methane-Stabilized Laser. *Phys. Rev. Lett.*, 29(19):1346–1349, 1972.

H. I. Ewen and E. M. Purcell. Observation of a Line in the Galactic Radio Spectrum. *Nature*, 168:356, 1951.

E. Feenberg. Conventionality in Distant Simultaneity. *Found. Phys.*, 4(1):121–126, 1974.

E. Fermi. Über die magnetischen Momente der Atomkerne. *Zeitschrift für Physik*, 60:320–333, 1930.

J. Finkelstein. Comment on "A one-way speed of light experiment" by E. D. Greaves, An Michel Rodriguez and J. Ruiz-Camacho [Am. J. Phys. 77 (10), 894-896 (2009)]. *Am. J. Phys.*, 78(8):877, 2010.

D. J. Fixsen, E. S. Cheng, J. M. Gales, J. C. Mather, R. A. Shafer, and E. I. Wright. The cosmic background spectrum from the full COBE FIRAS data set. *Astrophys. J.*, 473:576–587, 1996.

P. Frank and H. Rothe. Über die Transformation der Raumzeitkoordinaten von ruhenden und bewegten Systemen. *Ann. Physik*, 34(5):825–855, 1911.

S. J. Freedman and J. F. Clauser. Experimental test of local hidden-variable theories. *Phys. Rev. Lett.*, 28:938–941, 1972.

SF. Fung and K. C. Hsieh. Is the isotropy of the speed of light a convention? *Am. J. Phys.*, 48(8):654–657, 1980.

T. K. Gaisser, R. Engel, and E. Resconi. *Cosmic Rays and Particle Physics*. Cambridge University Press, 2016.

G. Gamow. Remarks on Lorentz contraction. *Proc. Nat. Acad. Sci (USA)*, 47:728–729, 1961.

G. Gamow. *Mr Tompkins in Paperback*. Cambridge University Press, 1965.

P. Gerber. Die räumliche und zeitliche Ausbreitung der Gravitation. *Zeitschrift für Mathematik und Physik*, 43: 93–104, 1898.

W. Gerlach and O. Stern. Das magnetische Moment des Silberatoms. *Zeitschrift für Physik*, 9:353–355, 1922.

C. N. Gordon. The Transverse Contraction Factor of Special Relativity. *Am. J. Phys.*, 40(5):782, 1972.

E. D. Greaves, A. M. Rodriguez, and J. Ruiz-Camacho. A one-way speed of light experiment. *Am. J. Phys.*, 77 (10):894–896, 2009.

T. Greber and H. Blatter. Aberration and Doppler shift: The cosmic background radiation and its rest frame. *Am. J. Phys*, 58(10):942–945, 1990.

T. Greber and H. Blatter. Intergalactic spaceflight: an uncommon way to relativistic kinematics and dynamics. *arXiv:physics/0608040*, 2006.

K. Greisen. End to the Cosmic-Ray Spectrum. *Phys. Rev. Lett.*, 16(17):748–751, 1966.

P. Gruner. Eine elementare Darstellung der Transformationsformeln der speziellen Relativitätstheorie. *Physikalische Zeitschrift*, 22:384–385, 1921.

J. Guala-Valverde, P. Mazzoni, and R. Achilles. The homopolar motor: A true relativistic engine. *Am. J. Phys.*, 70(10):1052–1055, 2002.

J. C. Hafele. Relativistic Behaviour of Moving Terrestrial Clocks. *Nature*, 227:270–271, 1970.

J. C. Hafele. Relativistic Time for Terrestrial Circumnavigations. *Am. J. Phys.*, 40(1):81–85, 1972.

J. C. Hafele and R. E. Keating. Around-the-World Atomic Clocks: Predicted Relativistic Time Gain. *Science*, 177: 166–168, 1972a.

J. C. Hafele and R. E. Keating. Around-the-World Atomic Clocks: Observed Relativistic Time Gain. *Science*, 177: 168–170, 1972b.

E. R. Harrison. The dark night-sky riddle: A "paradox" that resisted solution. *Science*, 236:941–945, 1984.

H. Hertz. Über die Einwirkung einer geradlinigen electrischen Schwingung auf eine benachbarte Strombahn. *Ann. Physik*, 270(5):155–170, 1888a.

H. Hertz. Über die Ausbreitungsgeschwindigkeit der electrodynamischen Wirkungen. *Ann. Physik*, 270(7): 551–569, 1888b.

B. Hoffmann. *Relativity and its Roots*. W.H. Freeman and Company, 1983.

E. Hubble. Effects of Red Shifts on the Distribution of Nebulae. *Astrophys. J.*, 84(12):517–554, 1936.

K. Hutter and K. Jöhnk. *Continuum Methods of Physical Modeling*. Springer Verlag, Berlin, Heidelberg, 2004.

W. Ignatowsky. Einige allgemeine Bemerkungen zum Relativitätsprinzip. *Phys. Zeitschr.*, 11(21):972–977, 1910.

J. Jackson. *Classical Electrodynamics, Second Edition*. John Wiley & Sons, Inc., 1975.

R. Jost. *Elektrodynamik, Nach einer Vorlesung an der ETH Zürich im Sommersemester 1975*. Verlag der Fachvereine an den Schweizerischen Hochschulen und Techniken, Zürich, 1976.

A. Komar. Foundations of Special Relativity and the Shape of the Big Dipper. *Am. J. Phys.*, 33(12):1024–1027, 1965.

Y. Kumazaki, T. Akagi, T. Yanou, D. Suga, Y. Hishikawa, and T. Teshima. Determination of the mean excitation energy of water from proton beam ranges. *Radiation Measurements*, 42:1683–1691, 2007.

C. Lagoute and E. Davoust. The interstellar traveller. *Am. J. Phys.*, 63(3):221–227, 1995.

W. E. Lamb and R. C. Retherford. Fine Structure of the Hydrogen Atom by a Microwave Method. *Phys. Rev.*, 72:241–243, 1947.

A. Lampa. Wie erscheint nach der Relativitätstheorie ein bewegter Stab einem ruhenden Beobachter? *Zeitschr. Phys.*, 27(1):138–148, 1924.

P. S. Laplace. *Exposition du Systéme du Monde, tome 2*. Imprimerie du Cercle Social, Paris, 1796.

J. M. Levy-Leblond. Speed(s). *Am. J. Phys.*, 48(5):345–347, 1980.

J. M. Levy-Leblond and J.-P. Provost. Additivity, rapidity, relativity. *Am. J. Phys.*, 47(12):1045–1049, 1979.

E. Loedel. Aberration y Reladividad. *Anales Sociedad Cientifica Argentina*, 145:3–13, 1948.

H. A. Lorentz. *Versuch einer Theorie der electrischen und optischen Erscheinungen in bewegten Körpern*. Verlag E.J. Brill, Leiden, 1895.

P. Lorrain, D. R. Corson, and F. Lorrain. *Electromagnetic Fields and Waves, Third Edition*. W. H. Freeman and Company, New York, 1988.

J. C. Maxwell. *A Treatise on Electricity and Magnetism*. Macmillan and Co, London, Publishers to the University of Oxford, 1873.

J. Michell. On the Means of discovering the Distance, Magnitude etc. of the Fixed Stars, in consequences of the Diminution of the velocity of their Light, in case such a Diminution should be found to take place in one of them, and such other Data should be procured from Observations, as would be farther necessary for that purpose. *Phil. Trans. Royal Soc. London*, 74:35–57, 1784.

A. A. Michelson and E. W. Morley. On the relative Motion of the Earth and the Luminiferous Ether. *Am. J. Sci.*, 134(203):333–345, 1887.

H. Minkowski. Raum und Zeit. *Jahresbericht der Deutschen Mathematiker-Vereinigung*, 18:75–88, 1909.

A. Mirabelli. The ether just fades away. *Am J. Phys.*, 53(5):493–494, 1985.

C. W. Misner, K. S. Thorne, and J. A. Wheeler. *Gravitation*. W. H. Freeman and Company, San Francisco, 1973.

T. Müller, A. King, and D. Adis. A trip to the end of the universe and the twin paradox. *Am. J. Phys.*, 76(4-5): 360–373, 2008.

P. J. Nahin. *Oliver Heaviside: the life, work, and times of an electrical genius of the Victorian age*. John Hopkins University Press, 1988.

T. Needham. *Visual complex Analysis*. Oxford University Press, 1997.

I. Newton. *Philosophiae Naturalis Principia Mathematica*. S. Pepys, Reg. Soc. Praeses, 1686.

C. Nissim-Sabat. Can one measure the one-way velocity of light? *Am. J. Phys.*, 50(6):533–536, 1982.

P. Øhrstrom. Conventionality in Distant Simultaneity. *Found. Phys.*, 10(3/4):333–343, 1980.

L. B. Okun. The Concept of Mass. *Physics Today*, 42:31–36, 1989.

L. A. Pars. The Lorentz Transformation. *Philos. Mag. Ser. 6*, 42(248):249–258, 1921.

T. J. Pearson, S. C. Unwin, M. H. Cohen, R. P. Linfield, A. C. S. Readhead, G. A. Seielstad, R. S. Simon, and R. C. Walker. Superluminal expansion of quasar 3C273. *Nature*, 290:365–368, 1981.

R. Penrose. The apparent shape of a relativistically moving sphere. *Proc. Cambridge Philos. Soc.*, 55:137–139, 1959.

H. Poincaré. Sur la dynamique de l'electron. *Comptes Rendues des Séances de l'Academie des Sciences*, 140: 1504–1508, 1905.

H. Poincaré. Sur la dynamique de l'electron. *Rendiconti del Circolo metematico di Palermo*, 21:129–176, 1906.

R. V. Pound and G. A. Rebka Jr. Gravitational Red-Shift in Nuclear Resonance. *Phys. Rev. Lett.*, 3(9):439–441, 1959.

G. Preti. Schwarzschild Radius Before General Relativity: Why Does Michell-Laplace Argument Provide the Correct Answer? *Found. Phys.*, 39:1046–1054, 2009.

C. V. Raman and S. Bhagavantam. Experimental proof of the spin of the photon. *Indian Journal of Physics*, 6: 353–366, 1931.

H. Reichenbach. *Axiomatization of the Theory of Relativity*. University of California Press, Berkeley, 1969.

G. Roddenberry. Star Trek: The Motion Picture (Film). Paramount Pictures, 1979.

M. Römer. Demonstration touchant le mouvement de la lumiere. *Des Scavans*, pages 233–236, 1676.

W. C. Röntgen. Über eine neue Art von Strahlen. *Sitzungsberichte der Würzburger Physik.-medic. Gesellschaft Würzburg*, pages 137–147, 1895.

A. R. Sandage, R. G. Kron, and M. S. Longair. *The Deep Universe*. Springer, New York, 1993.

J. Scalzi. *Old Man's War*. Tor Books, New York, 2005.

U. E. Schröder. *Gravitation, Einführung in die Allgemeine Relativitätstheorie*. Verlag Harri Deutsch, Thun, 2007.

K. Schwarzschild. Über das Gravitationsfeld eines Massenpunktes nach der Einsteinschen Theorie. *Sitzungsberichte der Königlich-Preussischen Akademie der Wissenschaften*, 1:189–196, 1916.

F. W. Sears. Tempest in the Relativistic Teapot. *Am. J. Phys.*, 34(4):363, 1966.

C. Semay and B. S. Silvestre-Brac. The equation of motion of an interstellar Bussard ramjet with radiation loss. *Acta Astron.*, 61:817–822, 2007.

C. Semay and B. S. Silvestre-Brac. The equation of motion of an interstellar Bussard ramjet with radiation and mass losses. *Eur. J. Phys.*, 29:1153–1163, 2008.

R. U. Sexl and H. K. Urbantke. *Gravitation und Kosmologie, eine Einführung in die Allgemeine Relativitätstheorie*. B.-I.-Wissenschaftsverlag, Bibliographisches Institut, Mannheim, Wien, Zürich, 1983.

M. A. Shupe. The Lorentz-invariant vacuum medium. *Am. J. Phys.*, 53(2):122–127, 1985.

D. N. Spergel et al. First-Year Wilkinson Microwave Anisotropy Probe (WMAP) Observations: Determination of Cosmological Parameters. *Astrophys. J. Suppl. Ser.*, 148:175–194, 2003.

L. Spitzer, Jr. *Physical Processes in the Interstellar Medium*. John Wiley & Sons, New York, 1978.

G. E. Stedman. A unidirectional test of special relativity. *Am. J. Phys.*, 40(5):782–784, 1972.

G. E. Stedmann. Reply to Ehrlichson:Is the Apparent Speed of Light Independent of the Sense in Which it Traverses a Closed Polygonal Path? *Am. J. Phys.*, 41(11):1300–1302, 1973.

E. F. Taylor and J. A. Wheeler. *Spacetime Physics*. W. H. Freeman and Company, San Francisco, 1963.

J. Terrell. Invisibility of the Lorentz Contraction. *Phys. Rev.*, 116(4):1041–1045, 1959.

L. H. Thomas. The kinematics of an electron with an axis. *Philosophical Magazine*, 7:1–23, 1927.

P. A. Tipler. *Physics: for scientists and engineers*. W.H. Freeman, New York, 1999.

P. Touboul. MICROSCOPE Mission: Final Results of the Test of the Equivalence Principle. *Phys. Rev. Lett.*, 129,121 102, 2022.

A. A. Ungar. The Lorentz transformation group of the special theory of relativity without Einstein's isotropy convention. *Philos. Sci.*, 53:395–402, 1986.

A. A. Ungar. Ether and the one-way speed of light. *Am. J. Phys*, 56(9):814, 1988.

A. A. Ungar. Thomas precession and its associated grouplike structure. *Am. J. Phys.*, 59:824–834, 1991a.

A. A. Ungar. Formalism to Deal with Reichenbach's Special Theory of Relativity. *Found. Phys.*, 21(6):691–726, 1991b.

A. C. Véron and T. Greber. Non-relativistic aberation and Doppler shift as experienced in walking with different velocities relative to falling rain (1.0). , 2024. URL https://doi.org/10.5281/zenodo.13983131.

I. Vetharaniam and G. E. Stedman. Synchronisation convention in test theories of special relativity. *Found. Phys. Lett.*, 4(3):275–281, 1991.

J. G. von Soldner. Ueber die Ablenkung eines Lichtstrals von seiner geradlinigen Bewegung, durch die Attraktion eines Weltkörpers, an welchem er nahe vorbei geht. *Berliner Astronomisches Jahrbuch*, pages 161–172, 1804.

H. Weinberger and M. Mossel. Theory for a Unidirectional Interferometric Test of Special Relativity. *Am. J. Phys.*, 39(6):606–609, 1971.

C. A. Will. Henry Cavendish, Johann von Soldner, and the deflection of light. *Am. J. Phys.*, 56(3):413–415, 1988.

R. R. Wilson. Radiological use of fast protons. *Radiology*, 47:487–491, 1946.

A. Yeghikyan and H. Fahr. Effects induced by the passage of the Sun through dense molecular clouds. *Astron. Astrophys*, 415:763–770, 2004.

G. T. Zatsepin and V. A. Kuzmin. Upper Limit of the Spectrum of Cosmic Rays. *Soviet Journal of Experimental and Theoretical Physics Letters*, 4:78, 1966.

E. C. Zeeman. Causality Implies the Lorentz Group. *J. Math. Phys.*, 5(4):490–493, 1964.

Index

https://doi.org/10.1515/9783111503592-011

Also of interest

Gravitation und Relativität.
Eine Einführung in die Allgemeine Relativitätstheorie
Holger Göbel, 2023
ISBN 978-3-11-120033-0,
e-ISBN 978-3-11-120240-2

Relativistic World. Volume 1: Mechanics
Sergey Stepanov, 2018
ISBN 978-3-11-120033-0,
e-ISBN 978-3-11-051588-6

Quantum Technologies. For Engineers
Rainer Müller, Franziska Greinert, 2023
ISBN 978-3-11-071744-0,
e-ISBN 978-3-11-071745-7

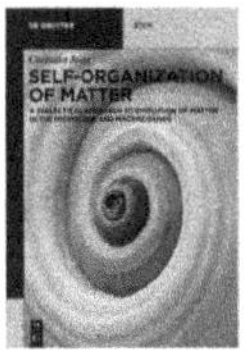

Self-organization of Matter.
A dialectical approach to evolution of matter in the microcosm and macrocosmos
Christian Jooss, 2020
ISBN 978-3-11-064419-7,
e-ISBN 978-3-11-064420-3